CHEMISTRY

IN

100

NUMBERS

A Numerical Guide to Facts, Formulas and Theories

JOEL LEVY

APPLE

Table of contents

CHEMISTRY

IN

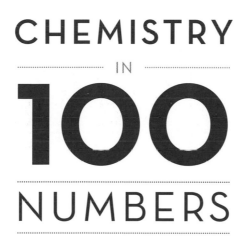

100

NUMBERS

A Numerical Guide to Facts,
Formulas and Theories

First published in the UK in 2015 by
Apple Press
74-77 White Lion Street
London N1 9PH

www.apple-press.com

ISBN: 978 1 84543 607 0

Conceived, designed, and produced by
Quid Publishing
Level 4 Sheridan House
114 Western Road
Hove BN3 1DD
England

Design and illustration by Simon Daley

Dedication: For Ruth, the best collection of atoms in the universe

www.quidpublishing.com

Printed in China

10 9 8 7 6 5 4 3 2 1

Introduction

'Number is the principle clue which leads to wisdom', wrote the 15th-century theologian and natural philosopher Nicholas of Cusa. Following Plato, Nicholas believed that the material world is only a shadow of the world of ideas, and the only way to glimpse this ideal world is through the prism of numbers. Galileo argued along similar lines with his famous credo that the Book of Nature is written in the language of numbers.

An insistence on the primacy of numbers might appear to be at odds with chemistry, the study of matter and its transformations, aka 'the science of stuff'. Chemistry may be at its most appealing when it resembles the alchemy of yore, as a pinch of powdered permanganate is mixed with a soupçon of solvent, or a crucible of calx is blasted with a Bunsen burner; who needs numbers when cooking up potions and poultices? But the absence of numbers lay at the heart of the failings and false starts of alchemy; numbers were precisely what transmogrified the magic and mysticism of alchemy into the science of chemistry. It was only when natural philosophers stopped obscuring their work in the language and symbolism of the esoteric, and started to quantify ingredients, treatments and techniques that their obscure art could become an enlightened science.

Chemistry in 100 Numbers seeks to reflect and illuminate the importance of numbers to the history of chemistry. Featured amongst this compendium of the most important and interesting numbers in chemistry are numbers that describe the earliest theories about matter, the origins of the laws of chemistry and the development of chemistry as a science. Fundamental constants of the natural world are featured, alongside the quantities most important for life on Earth. Key concepts in chemistry are explained in accessible terms, so that the entries build up into a complete introduction to the whole panoply of chemistry, from materials science to nuclear chemistry, from the periodic table to poisons.

Above all, however, the 100 entries of this book contain an (almost) innumerable host of fascinating trivia: which industrial process is the source of half of the protein in your body, and why there are some serious problems with the kilogram; why anything over 350 is a bad number for the global climate and Marie Curie's notebooks are too hot to handle; how to hold up a cat with a whisker, and how a roll of sticky tape helped two Anglo-Russian researchers win a Nobel Prize; how the ancient Greek philosophers were afflicted with element inflation; why diamond is not the hardest substance known, and how ancient Egyptian mummification technology is represented in the periodic table; why it's a myth that you can 'cook off' alcohol, and why you should blot your steaks on a paper towel before flash-frying them; how you can slow down light enough to outrun it, and why one of the elements is named 'smelly' while two others are named after types of goblin. With apologies to Galileo, this is a book of numbers written in the language of nature.

-92.4

Enthalpy of the Haber
process (kJ/mol)

92.4 is the amount of energy released in kilojoules (kJ) for every mole of ammonia formed in the Haber process (a mole is the amount of a substance consisting of an Avogadro's number of particles – see page 168). Enthalpy is the change in heat content of a system, and when a chemical reaction occurring in a system is exothermic, meaning that energy is released or given off, the enthalpy is said to be negative, hence the minus sign.

This book is not arranged in a linear way; each entry should be read separately and there should, in theory, be no significance to the fact that this is the first entry. In practice, however, it is highly fitting, because the Haber process (also commonly called the Haber-Bosch process – see below) is probably the single most important chemical discovery in human history in the sense that it has had the most impact on the greatest number of people. Even for you, the reader, this obscure chemical process is of colossal personal importance, because almost half of the nitrogen in your body is there because of it. Nitrogen is a key ingredient of protein, the class of biological molecules that makes up much of your body, including your muscles, enzymes, cell structures and many other bits.

Fixing nitrogen

The Haber-Bosch process is the way in which nitrogen is 'fixed' on an industrial scale. Because of its role in making proteins, nitrogen

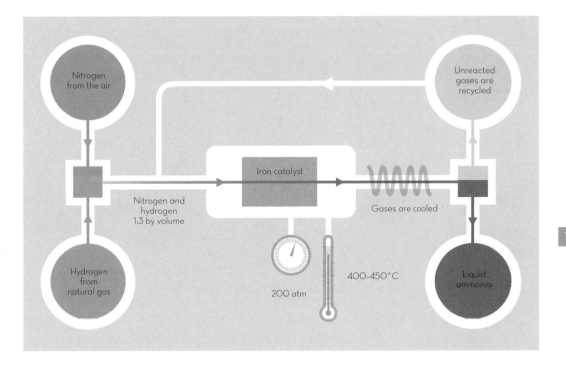

Nitrogen from the air

Unreacted gases are recycled

Iron catalyst

Nitrogen and hydrogen 1:3 by volume

Gases are cooled

Hydrogen from natural gas

200 atm

400-450°C

Liquid ammonia

is an essential ingredient for growing plants and animals, but although nitrogen is plentiful in the atmosphere, it is extremely unreactive and thus very hard to obtain during biological processes (such as plant growth). Fixing nitrogen means binding it to other elements to form compounds that are more biologically accessible, which can then be used to make fertiliser. In 1909, German chemist Fritz Haber worked out the conditions facilitating the process that now bears his name, showing that nitrogen gas and hydrogen gas would react under pressure and in the presence of a catalyst to produce ammonia (NH_3). The German industrial chemist Carl Bosch industrialised the process, making use of the negative enthalpy of the reaction (i.e. the fact that it produces heat) to help drive it by using it to heat the gases coming into the reaction chamber.

▲ Flow chart showing the main steps in the Haber process. In the intense heat and pressure of the reaction chamber, nitrogen and hydrogen react to give ammonia.

Explosive demands

Haber was driven to this breakthrough by military imperatives. Fixed nitrogen is essential for production of explosives as well as fertiliser, and Germany was gearing up for a European war, which was likely to see the British Navy blockade the maritime route for the supply of guano (bird droppings, the main natural source of fixed nitrogen at this point in history). During the First World War, the Haber-Bosch process enabled German industry to maintain its production of explosives despite the loss of the guano supply, keeping Germany in the war. After the war, the process would have an explosive impact on global agriculture and, by extension, the world population. Industrial synthesis of ammonia on a massive scale meant that the primary limiting factor for agriculture – lack of nitrogen – could now be overcome by using ammonia to make ammonium nitrate (NH_4NO_3) and other fertilisers. Global food production increased exponentially, as did global population, which increased fourfold by the end of the 20th century (see graph).

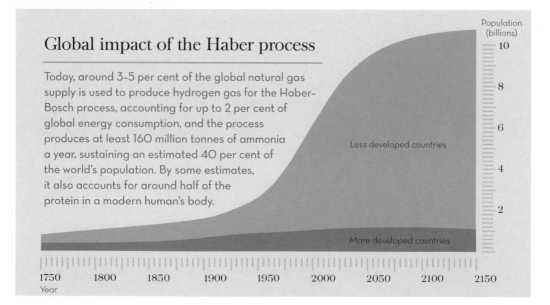

Global impact of the Haber process

Today, around 3-5 per cent of the global natural gas supply is used to produce hydrogen gas for the Haber-Bosch process, accounting for up to 2 per cent of global energy consumption, and the process produces at least 160 million tonnes of ammonia a year, sustaining an estimated 40 per cent of the world's population. By some estimates, it also accounts for around half of the protein in a modern human's body.

Less developed countries

More developed countries

Population (billions)

10

8

6

4

2

1750   1800   1850   1900   1950   2000   2050   2100   2150
Year

-90

Boiling point of water in the absence of H-bonds (°C)

90 below zero on the centigrade or Celsius scale (183 K) is the temperature at which water would change phase from liquid to gas (i.e. evaporate or boil), if it were not for the presence of H-bonds. The great disparity between this and the actual value (by definition, 100°C) indicates just how important this is to life on Earth.

The angular shape of the bonds in a water molecule leads to partial polarity of the molecule, effectively turning it into a sort of tiny magnet (see 104.5°, page 103). This in turn means that the partially positively charged hydrogen 'end' of one water molecule can form a weak ionic bond with the partially negatively charged oxygen 'end' of another. This bond is known as an H-bond (see also page 103).

In liquid water, the H_2O molecules are quite energetic. Each one is constantly making, breaking and re-forming H-bonds, like partners at a barn dance or ceilidh, except that water molecules change partners billions of times a second. At any one moment, only 15 per cent of the molecules in a glass of water are actually touching. But this means there are enough H-bonds present to alter dramatically the properties of liquid water. Generally, the boiling point of a liquid is proportional to its molecular weight, and substances with similar molecular weight to H_2O are usually gases at room temperature. If this were true of water, there would be almost no liquid water on Earth, making life impossible on most of the planet.

-3

Charge on phosphorus anion

14

A phosphorus atom that gains three electrons becomes an ion with a negative charge, known as an anion.

Atoms seek to obtain the most stable configuration of electrons in their outer shells (see 'valency' and 'the octet rule', pages 52 and 60), sharing, losing or gaining electrons as they seek to become isoelectronic with their nearest noble gas (see page 61). In a covalent bond, an atom shares electrons with another atom, partially donating or receiving electrons. But when electrons are completely donated or received, the atom ends up with a mismatch between the number of (positively charged) protons in the nucleus and (negatively charged) electrons orbiting the nucleus, and this leads to the whole atom becoming positively or negatively charged, at which point it is known as an ion.

▲ This is the scientific notation for a phosphorus anion with a charge of -3: the charge is shown by a superscript suffix to the standard abbreviation for the element.

Anions and cations

In scientific notation, the charge on an ion is written as a superscript following the chemical symbol. For instance, P^{-3} is the notation for the ion of phosphorus with a negative charge of 3, indicating that the phosphorus atom has gained three electrons. A negative ion is known as an anion, while a positively charged ion is known as a cation. For example, Na^+ is the cation formed by a sodium atom that has given up a single electron, so that it has 11 protons in the nucleus but only 10 electrons around the nucleus, giving it an overall positive charge of 1.

Ionic bonds

Ions with opposing charges experience an electrostatic attraction, known as an ionic bond. For instance, when a sodium atom (Na) and a chlorine atom (Cl) combine to form table salt (NaCl), what has actually happened is that the sodium has donated an electron to the chlorine, so that the compound might more properly be written Na^+Cl^-. The sodium cation and the chlorine anion are ionically bonded together. Metal salts such as NaCl typically form a crystalline lattice in their solid form. Because water molecules are polarised (see '104.5°', page 103), they easily form complexes with ions, and this means that ionic compounds such as salts are usually soluble in water.

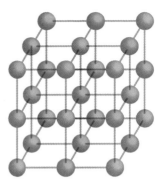

▲ Diagrammatic view showing the arrangement of sodium cations and chlorine anions in the lattice that makes up a crystal of table salt.

Periodic ion rules

P^{-3} is an example of a monatomic ion (it comprises just one atom). In monatomic ions, the number of electrons lost or gained, and hence the charge on the ion, is determined by periodic rules (i.e. the properties characteristic of the part of the periodic table in which the element sits). The number is usually the same as the valency and characteristic oxidation state of the element; all of them are dependent on the number of electrons in the outer shell of an atom of the element.

So, for instance, group I of the periodic table, the alkali metals, are elements with a single electron in the outer shell. To become isoelectronic with the nearest noble gas they tend to lose this electron easily, giving +1 cations; hence lithium, sodium and potassium all form cations: Li^+, Na^+ and K^+. In group 2, the alkaline earth metals, there are two electrons in the outer shell and when these are lost the resulting cation is therefore +2, hence magnesium and calcium give the ions Mg^{2+} and Ca^{2+} respectively.

9.109390×10^{-31}

Mass of the electron (kg)

An electron is a fundamental particle with an infinitesimal mass; to write out the mass of an electron in kilograms in full, without using notation, you would have to write a 0 followed by a decimal place followed by 30 more zeros and finally a 9.

Not surprisingly, a quantity this minute is incredibly hard to measure directly. Perhaps the nearest scientists have come to achieving this feat was in an experiment reported in the journal *Nature* in February 2014, where an electron bound to a carbon nucleus was held in a charged construct called a 'Penning trap', enabling its mass to be determined to unprecedented levels of accuracy, as 0.000548579909067 of an atomic mass unit.

Finding the e in e/m

Yet the mass of the electron has been known since 1909, when the American physicist Robert Millikan performed his oil-drop experiment to measure the charge of the electron. Knowing the charge of the electron revealed its mass because the key discovery in the history of this tiny particle had come six years earlier, when British physicist J.J. Thomson had discovered the charge (e) to mass (m) ratio of the electron to be 1.7588196×10^{11} coulomb/kg (see page 166). If $e/m = 1.7588196 \times 10^{11}$, then $m = e/1.7588196 \times 10^{11}$.

Millikan's oil-drop experiment is famous for several reasons: it was the first successful scientific attempt to detect and measure the effect of an individual subatomic particle, winning Millikan the 1923 Nobel Prize in Physics; it was noted for the elegance with which it was devised and carried out; it provided vital evidence for the quantum nature of subatomic physics; and it has become the source of enormous controversy with accusations of scientific fraud (see page 19).

The oil-drop experiment

Just as Thomson's experiment to measure the charge/mass ratio of the electron had ingeniously circumvented the difficulties in direct measurement of tiny particles by using balancing forces (see page 167), Millikan devised an apparatus that did something similar. He reasoned that if he could balance electrostatic attraction against gravity, using charged plates to stop charged droplets of oil from falling, he could find the strength of the charge on each droplet by determining the mass of the droplet (which would be relatively easy to work out). The apparatus Millikan used was an atomiser that sprayed tiny droplets of oil into an ionised

▼ Simplified diagram of Millikan's experiment, showing the drops created by the atomiser falling into the space between the charged plates, where the observer can use the microscope to determine when they stop falling.

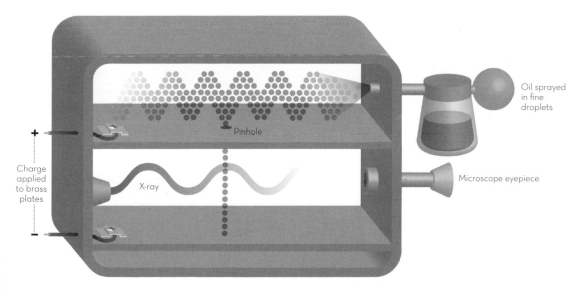

Oil sprayed in fine droplets

Pinhole

Charge applied to brass plates

X-ray

Microscope eyepiece

+

−

air space between two charged plates, which he could observe through a microscope. Spraying out droplets like this gives a good chance that they will pick up an electrical charge. Watching a droplet through the microscope, Millikan could adjust the strength of the electric field between the plates until he 'caught' the drop and it stopped falling. Since it was trivial to work out the mass of the droplet (by turning off the field and timing how fast the drop fell at terminal velocity, using this to determine the volume of the oil drop and multiplying the volume by the density of the oil), the strength of the field allowed Millikan to work out the charge on the oil drop.

Then Millikan blasted the space with ionising radiation (he used X-rays), which zaps electrons with enough energy to jump off or onto the oil drop. This changes the charge on the droplet, causing it to start moving towards or away from the positively charged plate; by measuring how much the strength of the electric field had to be changed to 'catch' the oil drop, Millikan was measuring the quantity of charge that had been added to or lost from the drop. He found that all the measurements were multiples of one number, currently accepted as 1.6×10^{-19} coulombs.

Quantum of charge

The most likely explanation for this finding is that each electron that is added to or lost from the oil drop has a certain fixed quantity of charge, and that the lowest common denominator, 1.6×10^{-19} C, is the charge on a single electron. This finding showed that the value of the charge on the electron is quantised, which in turn is vital evidence for the corpuscular or particle nature of electrons (i.e. proof that electrons are particles). At the time, many scientists believed they must be wave-like and not particle-like in nature; it has since been found that electrons, like light photons, are both wave-like and particle-like. More to the point, finding the charge (e) of the electron made it possible to solve the equation from above: $m = e/1.7588196 \times 10^{11} \rightarrow m = 1.6 \times 10^{-19}$ C$/1.7588196 \times 10^{11}$ C/kg $\sim 9 \times 10^{-31}$ kg.

Millikan the fraud?

The beauty and power of Millikan's experiment helped him to win the Nobel Prize, but his approach to recording the results of the experiment have since come in for close scrutiny. After his initial 1909 findings were challenged by another scientist who claimed to have shown the existence of 'sub-electrons', Millikan published a follow-up paper in 1913, in which he explicitly stated: 'This is not a selected group of drops, but represents all the drops experimented upon during 60 consecutive days.'

But inspection by historians of his original laboratory notebooks revealed that, for the period in question, he recorded data on many more drops than he eventually reported. Notes in the margin included comments such as 'Beauty, publish' and 'something wrong'. It has been suggested that, under pressure from a rival scientist, Millikan deliberately suppressed data that undermined his hypothesis and reported only observations that supported it, a common problem with scientific research believed to be a surprisingly widespread phenomenon among researchers. The controversy has dragged Millikan's name through the mud and his work is sometimes touted as a classic instance of scientific fraud.

A reassessment of the notebooks in question suggests that these accusations are false and Millikan did not intend or practise any deception. The nature of the experiment meant that many observations were unusable, or began well but went awry. For instance, for each drop Millikan would carry out multiple irradiations to change the charge, and then seek to 'recapture' the droplet, before letting it fall to measure its velocity. At any stage it was easy to lose the droplet, thus fouling the observations for that drop. Drops that were too big would fall too fast to measure, while those that were too small would be so buffeted by Brownian motion they could not be used. In other words, Millikan may have simply acted to present only meaningful observations, discounting those where the experiment had clearly not worked properly. Analysis of his results with the 'suppressed' data included have suggested that the determination of the elementary charge on the electron would not have been much affected anyway.

▲ Robert Andrews Millikan (1868-1953) won the Nobel Prize for his oil-drop experiments, but he also characterised and named cosmic rays and verified Einstein's equations relating to the photoelectric effect.

1.66×10^{-27}

Atomic mass unit (kg)

The atomic mass unit (amu) is $^1/_{12}$ of the mass of a nucleus of an atom of carbon-12. Measured in the SI unit of mass, the kilogram, 1 amu is 1.66×10^{-27} kg; the fact that this is not exactly the same as the mass of a proton or neutron, or even the average of a proton and neutron, is an important demonstration of Einstein's equation $E=mc^2$, proving that matter and energy are equivalent.

Atomic mass is not the same as atomic weight, although the two terms used to be used interchangeably and are still commonly mistaken. Weight is a force, which results from the acceleration of mass in a gravitational field – in other words, it is the heaviness of an object. Mass is a measure of the amount of matter in an object. The atomic mass of an element is given relative to a standard, which was set in 1961 as $^1/_{12}$ the mass of an atom of the isotope of carbon with 6 neutrons and 6 protons, ^{12}C or carbon-12. Because this measure is a ratio it has no units, so it is correct to say that the atomic mass of ^{12}C is simply 12. But to avoid confusion it is now common to use atomic mass units, abbreviated to amu or simply u (another name for this unit is the dalton, Da, named for John Dalton – see below).

A brief history of amu

The adoption of carbon-12 as the standard reference for amu is due to somewhat complicated historical reasons. John Dalton's atomic theory suggested that all the atoms of an element

▲ This is the scientific notation for an element (in this case lithium) along with important information, such as you might see in a periodic table. The indicated number below the element notation is the atomic mass of the element. Since this is a relative number it needs no units, although the term 'atomic mass units' is often used.

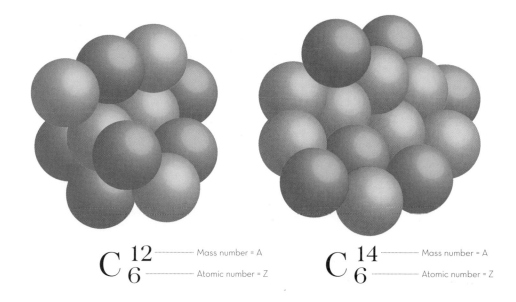

$$C^{12}_{6}$$ ⎯⎯⎯ Mass number = A
⎯⎯⎯ Atomic number = Z

$$C^{14}_{6}$$ ⎯⎯⎯ Mass number = A
⎯⎯⎯ Atomic number = Z

are identical and have the same mass, although back then chemists used the term 'weight' (for consistency, I will use the correct term throughout). Hydrogen was widely recognised as the lightest element, so it seemed natural to Dalton to assign to it a weight of 1 and define other elements in relation to it. But weighing individual atoms is clearly impossible (the mass of an atom of oxygen-16, for instance, is 2.657×10^{-23} grams, or 0.00000000000000000000002657 grams), and it was equally impossible for early chemists to say how many atoms or molecules were in their reagents. What they did know, thanks to Italian scientist Amedeo Avogadro, is that identical numbers of gas particles (atoms or molecules) occupy the same volume at a constant pressure and temperature, irrespective of their atomic mass. So a litre bag of oxygen contains exactly the same number of molecules as a litre bag of hydrogen. Comparing the weights of the two bags will reveal that oxygen is 16 times heavier than hydrogen. Since oxygen forms compounds with so many other elements, it was helpful for chemists to use oxygen as the reference standard and say that it had an atomic mass of 16, and in 1903 chemistry's International Committee on Atomic Weights set the amu as $1/16$ of the atomic mass of oxygen.

▲ Diagrammatic representations of the nuclei of two isotopes of the same element, carbon. The blue spheres represent protons, of which each isotope has six (hence both are atoms of carbon with the same atomic number [Z] and will be chemically though not physically identical). The grey spheres represent neutrons, with the carbon-14 atom on the right having two neutrons more than the carbon-12 atom, hence the difference in their mass numbers (A).

However, in 1919 it was discovered that, like many naturally occurring elements, oxygen exists as a mixture of isotopes, atoms with slightly different numbers of neutrons in the nucleus. So although the most common isotope is oyxgen-16, with 8 neutrons and 8 protons, naturally occurring oxygen also contains a small proportion of oxygen-17 and 18, with 9 and 10 neutrons respectively. This did not matter for chemists – as long as everyone was using the same mixture of oxygen isotopes in their experiments, which inevitably they were, 16 could stand as the atomic mass for the mixture of isotopes. Problems arose because physicists, dealing with individual atoms, did need greater specificity. Accordingly, physicists took the atomic mass of oxygen to refer to oxygen-16 only, leaving the scientific community with diverging definitions of an amu: one based on $\frac{1}{16}$ of the average mass of the oxygen atoms used by chemists, and the other based on $\frac{1}{16}$ of the mass of an atom of a particular isotope of oxygen.

▲ In the late 1950s the American physicist Alfred Nier (above left), the German physicist Josef Mattauch (above right) and the American chemist Edward Wichers were instrumental in convincing the scientific community to adopt carbon-12 as the reference standard for atomic mass.

This clash needed to be resolved, but if chemists adopted the physicists' definition, all their atomic weights would be off by a relatively large factor of 275 parts per million. A compromise was suggested: physicists already used the isotope carbon-12 as a standard in mass spectroscopy; if this were adopted as the reference for amu, chemists would only need to change their atomic masses by 42 parts per million, with a similar change for physicists. In 1961 the amu was accordingly set as $\frac{1}{12}$ the mass of a carbon-12 nucleus. Despite the compromise, the change nonetheless required some significant rewriting. For instance, pre-1961, the molecular weight of table salt, NaCl, was given as 58.45; after 1961 it became 58.44 – a difference of 0.02 per cent.

Missing mass

The nucleus of a carbon-12 atom contains 12 nuclear particles, so it might be expected that the mass of a single such particle in amu would be 1, and therefore that the atomic mass of hydrogen, which has a nucleus consisting of a single proton, should be 1. In fact, the atomic mass of hydrogen is 1.00794, which is more than $\frac{1}{12}$ of the atomic mass of carbon-12. Yet carbon-12 has 6 neutrons, which are a little bit more massive than protons; shouldn't it be heavier than 12 hydrogen nuclei? The discrepancy in mass (sometimes called the mass defect or deficit) arises because some of the mass in the carbon nucleus has been transformed into the energy needed to bind together the nuclear particles. This nuclear-binding energy has to be very powerful to overcome the electrostatic repulsion the positively charged protons have for one another. The precise amount of mass that needs to be converted into nuclear binding energy is governed by Einstein's famous equation, $E=mc^2$. The larger the nucleus, the more energy is needed. Changing the size of the nucleus changes the total mass of the system, resulting in the huge amounts of energy released in nuclear reactions.

1×10^{-15}

Dilution factor of a
homeopathic remedy
(part per quadrillion)

1×10^{-15}, or 0.000000000000001, is a fraction equivalent to 1 part per quadrillion. This is the factor by which the 'active ingredient' of a homeopathic remedy is diluted to achieve a mildly 'potent' preparation.

At these concentrations, the amount of active ingredient present in each drop of homeopathic medicine is equivalent to one human hair out of all the hair on all the heads of all the people in the world. More potent preparations are diluted by even greater factors.

Same suffering

Homeopathy is a system of alternative medicine based on the work of 19th-century German physician Samuel Hahnemann. He was unimpressed with the prevailing beliefs of the medical profession, which was based on the ancient system of balancing the humours, where symptoms would be countered with their opposites to restore balance to the patient's physiology. Instead, he believed that treatment should involve mimicking the disease symptoms (hence homeo- ['same'] -pathy ['suffering']), but in drastically reduced form, stimulating the vital forces of the body into action to restore balance to the system. Thus fever might be treated with tiny doses of a herbal extract or animal poison known to cause fever in high doses.

▲ Samuel Hahnemann was a German physician of the 18th and 19th centuries who viewed contemporary medical practices as 'murderous' and developed an alternative approach that became homeopathy.

Hahnemann believed that tiny doses of the medicine could be dynamised or increased in potency through succussion (shaking and banging), and that his remedies obeyed a mystical principle known as the Law of Infinitesimals. Simply stated, this holds that the smaller the dose of the remedy, the more potent its effects. Accordingly, a base remedy, such as an alcoholic herbal tincture, is diluted by mixing one part with ten or a hundred parts of water. This dilution is repeated multiple times. A 7c dilution has been diluted by a factor of 100 seven times. Hahnemann himself recommended using 30c dilutions most of the time, which would mean the active ingredient would be present at a concentration of less than 1 part in 10^{60}.

Immaterial

This raises a serious problem for believers in homeopathy. The greatest dilution that is likely to contain at least one molecule of the original substance is 12c, so the preparations recommended by Hahnemann and still prescribed by modern homeopaths are statistically unlikely to contain a single molecule of the supposed active principle. In other words, they are just water (or sugar, in the case of a homeopathic pill). How can they possibly have any physiological, let alone therapeutic, effect?

Hahnemann devised his system before there was any understanding of molecules and atoms. He believed that the process of succussion released 'immaterial and spiritual powers', making substances more active. Presumably modern homeopaths believe these 'powers' can imbue water with special properties. In fact, this is precisely what one advocate for homeopathy claimed to have found, in a notorious study by Jacques Benveniste published in the journal *Nature* in 1988. Benveniste claimed to have proved that water could somehow retain a memory of the homeopathic substance originally added. If water can 'remember' the remedy, however, why not sewage as well? Benveniste's study was subsequently demolished, and homeopathy has repeatedly failed to demonstrate, in proper clinical trials, any therapeutic value above that of a placebo.

▼ Homeopathic tinctures of yore. Homeopaths might argue that they ought to be stored in opaque or smoked-glass bottles to prevent light degrading the potency of the contents; chemists would argue it makes little difference since they contain no active ingredients anyway.

0.529×10^{-10}

Bohr radius (m)

0.529×10^{-10} metres, or 0.0000000529 millimetres, is the Bohr radius, the distance between a hydrogen nucleus and the electron that orbits it in its lowest energy configuration, and hence the smallest size possible for a hydrogen atom.

The Bohr radius is a quantity that was worked out by Danish physicist Niels Bohr, who in 1911 was a young graduate student fresh from Copenhagen, come to work under J.J. Thomson at Cambridge University's Cavendish Laboratory. Bohr had come to help probe the structure of the atom, but he wanted to incorporate fresh ideas, with which Thomson was uncomfortable. The work of Planck and Einstein on the nature of light had undermined the old notion of light as continuous waves, introducing a quantised concept of light as discrete packets of energy (i.e. photons): analogue giving way to digital. Bohr wanted to introduce similar concepts of quantisation to the energy of the electron, but Thomson was not supportive, so Bohr moved to Manchester to work under Ernest Rutherford.

▲ Niels Bohr was a Danish physicist who helped found the discipline of quantum physics and was the man who brought news of the discovery of nuclear fission to America (see page 118).

Digital electrons

There he absorbed Rutherford's new 'orbital' model of the atom, in which the electron orbits the central nucleus like a planet orbiting the Sun. Under the classical, 'analogue' conception of energy, the electron could orbit at any distance, depending on how much energy it had. But this failed to answer the question of why an electron with this degree of freedom did not simply get attracted into the opposingly charged nucleus. Bohr thought that the quantum, 'digital' conception of energy might hold the answer: the electron could only have certain fixed energy levels, and hence could only occupy one of a number of fixed orbits. In 1913, Bohr took the simplest atomic system possible, the hydrogen atom, and looked at the lowest possible energy state of the electron in this system, known as the ground state. Incorporating elements of quantum physics, he was able to calculate the minimum possible radius of the orbit of this ground-state electron, arriving at a figure of 0.529×10^{-10} metres. Using metres as the scale for such an infinitesimal quantity is unwieldy, so scientists use a unit called the ångström, or Å: $1 \text{ Å} = 1 \times 10^{-10}$ m, so the Bohr radius is 0.529 Å.

▲ The Bohr model of the ground state of a hydrogen atom, showing the electron orbiting the nucleus at the Bohr radius.

Muonic hydrogen

This is the closest distance at which an electron can orbit a proton, and hence it gives a diameter of 1.58 Å as the smallest possible size for the hydrogen atom, the smallest conventional atom. If the electron were more massive, the Bohr radius would be smaller, which has been experimentally shown by replacing the electron with a muon, a similar particle that is 206.77 times heavier than an electron, to make muonic hydrogen, which is about 206.77 times smaller.

It is important to note that the Rutherford–Bohr model of the atom has now been superseded: electrons are not discrete particles orbiting at fixed radii, but fuzzy clouds of probability spread out across orbital zones. But the Bohr radius continues to be a useful tool for describing the minimum possible radius an electron can have. This is important for chemistry because it is electrons and their distances from the nucleus that determine the chemistry of an element.

1×10^{-10}

Lowest temperature in the history of the universe (K)

1×10-10 or 0.0000000001 K (aka 100 picokelvin, or 0.1 billionths of a K) is the lowest temperature ever achieved, probably in the history of the universe, by the Low Temperature Laboratory at the Helsinki University of Technology.

In 1848, William Thomson, Lord Kelvin, devised a temperature scale that would set as zero the coldest conceivable state, one in which there is no energy whatsoever and so no movement or vibration of particles. This null state, which is purely theoretical since it cannot exist in nature, is designated 'absolute zero'. Kelvin used centigrades as the degrees in his scale, but this risked confusion between absolute zero and 0°C, the freezing point of water, so in the 20th century the units for the Kelvin scale were renamed kelvins (K). An increment of 1 K is thus the same as 1°C, but 0 K is -273.15°C.

Although absolute zero is impossible to achieve, there are parts of the universe that are very cold. The average temperature of the universe, as determined by measuring the cosmic microwave background radiation (CMB), is just 2.73 K, while the coldest place in space is the Boomerang Nebula, where massive gas emissions have cooled the temperature to just 1 K. In 1999, using multiple cooling techniques, a team of researchers at Helsinki's Low Temperature Lab succeeded in cooling a piece of rhodium metal to 0.1 billionths of a kelvin, setting a world record. Because such extreme cooling does not occur in nature, it is possible that this is the lowest temperature achieved in the history of the universe.

5×10^{-7}

Amount of mass lost per year by the international prototype kilogram (g)

5×10^{-7} grams or 0.5 micrograms is the annual change in mass between the international prototype kilogram (IPK) and official copies, since the adoption of the international prototype in 1889.

The IPK is a cylinder of platinum–iridium alloy in a 90-10 mix, adopted in 1889 as the definition of the kilogram in the Système International d'Unités (international system of units, or SI). Most of the other base SI units are defined with regard to fundamental constants, such as the speed of light, but the kilogram is still defined with regard to this artefact. Identical copies of the IPK are distributed to national weights and measures authorities, and a number of them are stored alongside the IPK at the International Bureau of Weights and Measures (known by its French acronym, BIPM), at Sèvres in France.

Weight watchers

Platinum was chosen as the main metal for the IPK because it is dense and resistant to chemical, magnetic and static electrical corrosion; iridium was added to harden the soft platinum. Periodically the copies and national

▼ Graph showing the minute fluctuations and divergence over the last hundred years of the mass of the standard copies, in relation to the IPK.

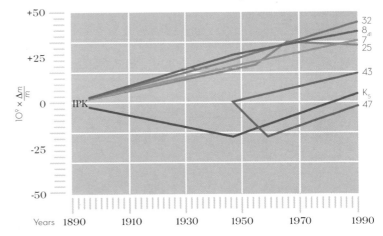

prototypes are checked against the IPK to ensure they still match, using a comparator balance. On three occasions between the end of the Second World War and 1990, the IPK and some of its copies were compared on a super-sensitive balance.

This programme of checking uncovered the peculiar anomaly that there has been a steady and consistent divergence in the mass of the IPK and its copies, to the tune of about 0.5 micrograms a year. It is impossible to say if the IPK is losing mass or the copies are gaining it. The cause of the change remains a mystery; although the BIPM notes suggestions that catalytic reactions between platinum and atmospheric substances might explain the phenomenon. It also points out that the IPK and several of its copies, which are of identical composition, are stored together and so must experience the same environmental factors.

Grave consideration

A change in mass of 50 micrograms over a hundred years is of little practical consequence. However, having a base SI unit based on an artefact rather than a fundamental constant is undesirable purely on grounds of scientific aesthetics, and becomes still more undesirable when the artefact exhibits mysterious phenomena. Accordingly, rather than wasting time exploring why the IPK is changing mass, the BIPM supports research aimed at linking the kilogram with fundamental constants so that the SI basis for its definition can be changed. Even once this is achieved, however, the SI unit of mass will remain aesthetically displeasing to particularly pedantic scientists because it is a base unit comprised of a thousand derivative units (despite being called the kilogram, it is the gram that is defined in relation to the kilogram, and not vice versa). Properly speaking, when it was adopted as an SI base unit the kilogram should have been renamed; one suggestion back in the 18th century was the *grave* (in the French sense of 'weighty').

1.00×10^{-7}

Concentration of hydroxonium ions present in pure water
(mol/dm^3)

1.00×10^{-7} moles per cubic decimetre is the concentration of hydroxonium (H_3O^+) ions in pure water. On the pH scale that measures the strength of acids and bases, this translates into 7, which is the exact mid-point.

Water naturally dissociates, so that in a pair of H_2O molecules one will act as a base and the other as an acid, with the former accepting a hydrogen ion (i.e. a proton) from the latter. This can be written as the equation:

$$H_2O + H_2O \rightarrow H_3O^+ + OH^-$$

The products of this reaction are a hydroxonium ion, a powerful acid, and a hydroxide ion, a powerful base. As these are so potent they rapidly neutralise each other to produce H_2O, so that the system is in equilibrium:

$$H_2O + H_2O \leftrightarrow H_3O^+ + OH^-$$

At room temperature, this equilibrium settles to produce a hydroxonium concentration of 1×10^{-7} mol/dm³. Since the reaction can be written in a simpler form as:

$$H_2O \rightarrow H^+ + OH^-$$

we can also say that the concentration of H^+ ions is 1.00×10^{-7}, and this equates to a pH of 7. So at room temperature, pure water has a pH of 7, and this is set as the neutral value for pH.

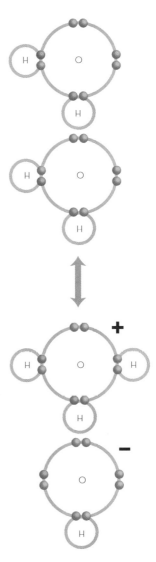

▲ Lewis diagram (see page 61) of water molecules showing how protons can be donated within pairs to give hydronium and hydroxide ions.

0.00008988

Density of hydrogen (g/cm^3)

Hydrogen has a density of just under 100 micrograms per cubic centimetre at standard temperature and pressure, making it extremely buoyant when compared to the density of the atmosphere.

This comparison measure, sometimes called the specific gravity of the gas, was first determined by the man who discovered hydrogen in 1766, Henry Cavendish, although he was neither the first to make hydrogen nor the one who named it. Hydrogen had been made as early as 1671 by Anglo-Irish natural philosopher Robert Boyle, who mixed filings of 'Mars' (iron) with dilute acid to evolve gaseous hydrogen, calling the mixture 'inflammable solution of Mars', but failing to identify hydrogen as a new element.

▲ The first manned hydrogen balloon flight, ascending from the Jardin des Tuileries in Paris on 1 December 1783.

Inflammable air

Nearly a century later, eccentric English nobleman Henry Cavendish was inspired to research the new field of pneumatic chemistry after reading of Scottish scientist Joseph Black's isolation of 'fixed air' (carbon dioxide). Cavendish evolved hydrogen in similar fashion to Boyle, passing the 'air' given off by the reaction through mercury and collecting it in an upturned vessel, an apparatus called a pneumatic trough. The new 'air' was highly inflammable, and he accordingly named it 'inflammable air', incorrectly surmising it was given off by the metals when in fact it came from the acid. Cavendish measured the amount of water that the new gas could displace, discovering that it had a specific gravity (density compared to the atmosphere) just $1/13$ that of common air, making it the lightest substance yet found.

Cavendish subscribed to the phlogiston theory, which held that combustion involved release of a fire-like element, phlogiston. He suspected that his inflammable air might be phlogiston itself, but this theoretical cul-de-sac did not prevent him from making a discovery that overturned millennia of thinking. In 1781, mixing hydrogen with oxygen (or inflammable with dephlogisticated air, in Cavendish's terminology) in a sealed vessel and lighting it with an electric spark, he produced water droplets, proving that water was a compound, not one of the primal elements, as had been believed since the time of the ancient Greeks (see page 54).

▲ The disastrous fate of the Hindenburg, which shattered public faith in hydrogen dirigibles and marked the end of the airship era.

33

Water maker

The phlogiston doctrine would be finally destroyed by the French chemist Antoine Lavoisier (see page 136), who showed that it had a back-to-front conception of combustion and related processes. Lavoisier devised new names for some of the recently discovered 'airs', christening Cavendish's inflammable air 'hydrogen', from Greek roots meaning 'water maker', following on from Cavendish's demonstration. Lavoisier adopted classical roots for his nomenclature to overcome linguistic borders, but in Germany, for instance, hydrogen is still known by its vernacular name *wasserstoff*, 'water stuff'.

The low density of hydrogen means that it is extremely buoyant in atmospheric air, and this did not escape the notice of the early balloon pioneers. Although the first balloon flight was achieved by the Montgolfier brothers in a hot air balloon, soon after Dr Jacques Alexandre Charles co-piloted the first manned hydrogen balloon, in December 1783. Hydrogen balloons could carry more and ascend higher and faster than hot-air ones, but the flammability of hydrogen made such balloons inherently dangerous, as famously demonstrated by the Hindenburg disaster of 1937.

0.025

LD_{50} of the venom of the inland taipan, the world's most venomous snake (mg/kg)

The dose at which the venom of the inland taipan snake is lethal to 50 per cent of mice injected just below the skin is 0.025 milligrams per kilogram. This means that for a mouse weighing about 20 grams, just 0.0005 milligrams of venom brings a 50 per cent chance of death.

All substances are poisons

Toxicity is a matter of degree. As Paracelsus famously observed, 'What is there that is not a poison? All substances are poisons, there is none that is not a poison. Only the dose determines that a thing is not a poison.' Even water is toxic in sufficiently large quantities. Table salt is not normally considered a poison, yet two tablespoons of salt is fatal to a toddler. How come? Toxicity depends on concentration, and this in turn will depend on body mass; toddlers are vulnerable to poisons because they are relatively small, so less is needed to produce harmful concentrations and thus harmful effects. Accordingly, toxic doses are given in units of milligrams/kilogram. Since 'anything can be a poison' is an unhelpfully broad definition, for legal purposes – in the US, at any rate – a poison is legally defined as a substance that is lethal at doses of 50 milligrams per kilogram of body weight or less. For an average adult male weighing around 85 kilograms, this means that a poison is a substance of which roughly a teaspoon's worth or less is fatal.

Hard on mice

Animal venoms are tested for lethality by injecting mice with different doses. The standard metric used in toxicology is the LD_{50}, or lethal dose-50: the dose at which 50 per cent of treated animals die. This figure will vary by administration route; the method of administration that most closely resembles a snake bite is subcutaneous injection. Measured on this basis, the snake with the most toxic venom is the inland taipan, *Oxyuranus microlepidotus*. The inland taipan can be up to 2 metres long but is generally timid and docile, feeding on small rodents and rarely biting humans. But the snake is far from harmless; with an LD_{50} of 0.025 mg/kg, 50 times more toxic than a king cobra, a single bite from an inland taipan contains enough venom to kill 100 human adults or up to 250,000 mice.

▼ Unlike many venomous animals, the inland taipan eschews garish warning colours and blends in with the arid surroundings of its typical habitat.

World's most dangerous snakes

This does not mean that the inland taipan is the most dangerous snake in the world. In fact, no deaths from its bite have ever been recorded, thanks to the availability of antivenin. The degree of danger a snake poses to humans depends on a wide variety of factors: how aggressive is the species, how much venom does it inject with each bite, how many bites will it inflict in a typical attack? These factors vary between individual snakes: younger snakes typically inject more venom than older ones. Even more important are human factors: how often are humans likely to come into contact with the snake, and to what extent is human development encroaching on the snake's habitat? How widely available is healthcare and, specifically, antivenins? These factors determine that snakes in the Indian subcontinent kill around 50–70,000 people a year, while in North America fewer than 20 people a year are killed.

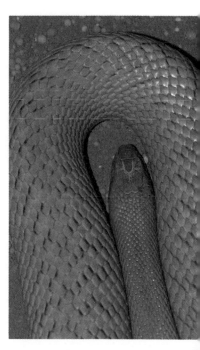

0.04

Proportion of atmosphere that is carbon dioxide (%)

0.04 per cent is the abundance or concentration of carbon dioxide in the atmosphere, although the precise figure fluctuates over time and is steadily increasing owing to anthropogenic (human-created) emissions.

Carbon dioxide (CO_2) was first tentatively identified by Flemish nobleman and alchemist Jan Baptista van Helmont in the early 17th century, but is said to have been properly discovered by Scottish physician and scientist Joseph Black in the 1750s. Van Helmont noted that burning charcoal released a new 'air', which he termed *gas* or *spiritus sylvester* ('spirit of the forest'); Black proved the existence of this new 'air' by careful weighing, burning *magnesia alba* (hydrated magnesium carbonate) to release carbon dioxide and showing that the surrounding air had gained mass while the *magnesia alba* had lost it. Because it had been bound up in a solid compound, Black called the new gas 'fixed air', and he surmised that it was a component of the atmosphere.

Vegetable measure

It became apparent that carbon dioxide was involved in the process now known as photosynthesis (in which plants take in carbon dioxide from the atmosphere and release oxygen), and in 1800 German polymath Alexander von Humboldt used this knowledge to make the first crude estimate of the atmospheric abundance of carbon dioxide. He sealed plants in a closed

▲ Around 1750 Scottish physician and scientist Joseph Black (above) made his name with a series of elegant and inspired demonstrations of the existence and nature of carbon dioxide, the gas first identified by the Dutch alchemist Jan Baptista van Helmont (below) over a century earlier.

chamber and measured the fall in volume of the air in the chamber as the plants took in CO_2 and grew, arriving at an estimate of 1 per cent for the concentration of carbon dioxide in the atmosphere. A similar methodology was used with far greater accuracy in 1804 by Theodore de Saussure, who carefully weighed the amount of organic matter and oxygen produced by plants in a sealed system, finding a value of around 0.04 per cent atmospheric CO_2 by volume.

The greenhouse effect

Since the late 20th century, this previously obscure measurement has become one of the most famous and controversial quantities in all of science, thanks to the role of carbon dioxide in global warming. Carbon dioxide is a 'greenhouse gas', so-called because it acts like the glass panels in a greenhouse. Greenhouse glass is transparent to incoming light but traps re-radiated heat, so warming sunlight can get into the greenhouse but heat cannot get out. Similarly, greenhouse gases are mostly transparent to visible light wavelengths of incoming solar radiation, letting it through to the lower atmosphere and the surface of the Earth. Here the solar radiation is absorbed, producing warming, which in return causes heat to be re-radiated as long-wave infrared radiation. Greenhouse gases absorb this long-wave radiation so it cannot escape into space, but is kept in the terrestrial ecosphere. The greenhouse effect caused by carbon dioxide, water vapour and other greenhouse gases is responsible for life on Earth as we know it, because without it the surface temperature of the planet would be around 35°C cooler.

Proxy climate records preserved in ancient ice, sediments, tree rings and other sources have allowed scientists to reconstruct past climates and atmospheric CO_2 levels, revealing a link between elevated levels and elevated temperatures. The overwhelming consensus of scientific opinion is that human activities – primarily the burning of fossil fuels for energy over the last 250 years or so – have caused a rapid and accelerating rise in atmospheric CO_2 levels. Powerful and well-funded lobbies

▲ The extraordinary Prussian explorer and naturalist Alexander von Humboldt was noted for his 'quantitative methodology', or emphasis on careful measurement, as exhibited in his early attempts to determine the concentration of carbon dioxide in the atmosphere.

oppose the scientific and economic corollaries of this conclusion, and so measurements of atmospheric carbon dioxide levels have become the focus of intense scrutiny and debate.

PPM rising

The key data source in this controversy has been the long-term CO_2 measurements begun by Dave Keeling of the Scripps Institution of Oceanography (SIO) in 1957, when he first collected a flask of air from the South Pole. The following year, the collection of air samples began on top of a mountain in Hawaii, and it is the concentration of CO_2 measured at the Mauna Loa Observatory in Hawaii that provides headline figures today. Because the abundance of CO_2 in the atmosphere is relatively small, using units of parts per million (ppm) rather than per cent gives much finer resolution for the measurements. The air that was sampled at Mauna Loa in March 1958 had a CO_2 concentration of 316 ppm. On 10 May, 2013, the US National Oceanic and Atmospheric Administration and the SIO first reported daily averages that temporarily reached 400 ppm. At the time of writing the level is 399 ppm – by the time you read this it will probably be over 400 ppm.

Heading for Venus?

The CO_2 level for most of human civilisation was around 275 ppm, but the current rate of increase is around 2 ppm per year, due to anthropogenic emissions. Consensus scientific opinion is that such a large and rapid rise in greenhouse gas abundance will cause profound changes to global climate and ecology, with serious and even catastrophic consequences for human populations and global biodiversity, including melting of ice caps and glaciers, rising sea levels, increased flooding, more severe storms and hurricanes, more heatwaves, more intense rainfall and consequent mudslides, drought in many regions, and deleterious effects on agriculture in many regions. The consensus is that a 'safe' level of CO_2 in the atmosphere is 350 ppm, so the gap

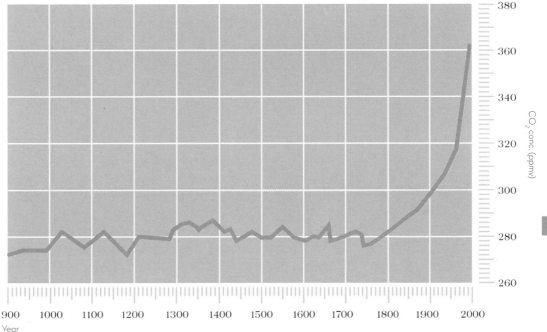

CO$_2$ conc. (ppmv)

Year

between this aspiration and the current level of 400 and rising causes widespread consternation.

Other planets in the solar system provide graphic illustrations of the consequences of both under- and over-abundance of CO$_2$. Mars, which has lost most of its atmosphere and has no tectonic activity to recycle carbon into the atmosphere, has very little atmospheric CO$_2$ and hence is extremely cold and apparently lifeless on the surface. Venus, which has a thick atmosphere, has very high levels of CO$_2$ and other greenhouse gases in the atmosphere, which have caused runaway warming and made the planet the hottest in the solar system, with surface temperatures of 462°C, hot enough to melt lead. This is even higher than temperatures on Mercury, the planet closest to the Sun.

▲ Proxy records of atmospheric CO$_2$ levels, combined with more recent direct measurements, make it possible to reconstruct how the atmospheric concentration of greenhouse gases has increased exponentially since the start of the Industrial Revolution in the late 18th century.

0.14

Radium in a tonne
of pitchblende (g)

0.14 g is the amount of the element radium present in a tonne of pitchblende (uranium ore). In order to isolate and characterise (i.e. discover) even a tiny amount of this new element, Marie Curie had to process tonnes of uranium ore.

In January 1896, the French scientist Henri Becquerel discovered that uranium salts emitted an invisible ray that could penetrate thick paper to fog a photographic plate. His discovery drew little attention in the scientific world; X-rays, discovered just a month earlier, were far more popular. But Becquerel's work did catch the eye of a brilliant young physicist looking for a subject for her doctoral thesis: Marie Curie. Born in Poland as Marie Sklodowska, she had overcome financial barriers and patriarchal attitudes to graduate at the top of her class from the Sorbonne in Paris. Shortly afterwards she met and married Pierre Curie, a dedicated and brilliant scientist with little in the way of money or connections. By 1898 Marie was looking to start a PhD, and thought that Becquerel's 'uranium rays' might be promising.

▲ Henri Becquerel (1852–1908) inherited an interest in – and supply of – uranium salts from his father.

In search of radium

Almost immediately she found that the mysterious rays were also given off by thorium. She and her husband termed this new form of energy 'radioactivity'. Of the known elements, only uranium and thorium exhibited radioactivity, but in a flash of inspiration Marie thought to investigate uranium ore, pitchblende. Interestingly, she found that it was far more radioactive than pure uranium, suggesting that it might contain as yet undiscovered radioactive elements. Further research produced evidence that present in minute amounts in the pitchblende was a highly radioactive element similar to bismuth, which the Curies named polonium, and one similar to barium, which they christened radium.

Isolating even minute amounts of these new elements would involve laborious processing of pitchblende, but in order to characterise the new element and formally identify it, Marie would need to obtain at least 100 milligrams. But she had an additional problem – she and her husband could not afford the expensive pitchblende ore to use as raw material. A clever workaround was reached when the mining company that produced pitchblende at the Joachimsthal mine in Bohemia agreed to donate several tonnes of the waste slag dumped in the forest around the mine after processing of the ore. Marie literally had to clear pine needles and other forest debris from the material before beginning the arduous, physically draining work of separation. 'Sometimes I had to spend a whole day stirring a boiling mass with a heavy iron rod nearly as big as myself. I would be broken with fatigue at day's end,' she later recorded.

The greatest thesis

After years of processing of tonnes of slag, Marie isolated 0.1 g of pure radium chloride, enabling her to determine its atomic weight as 225. When she presented her doctoral thesis in 1903, the doctoral committee (which included two future Nobel laureates) said that her discoveries represented the greatest scientific contribution ever made in a doctoral thesis. That same year she won the Nobel Prize in Physics (she would win another one, in Chemistry, eight years later). But the herculean effort involved would prove devastating to her health, owing to the prolonged exposure to radioactivity. Even today the Curies' laboratory notebooks are contaminated with radioactive material, and visitors to the Bibliothèque Nationale wishing to consult them have to sign a waiver acknowledging that they do so at their own risk.

▼ Marie Curie (shown here with her husband Pierre in their laboratory), has achieved legendary status as a trail blazer for women in a male-dominated field, a supremely dedicated researcher who gave her life for her work and, above all, a brilliant scientist who achieved the unique distinction of winning Nobel prizes in two different sciences.

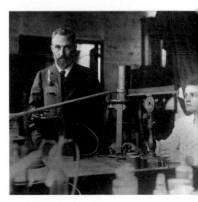

1

Thickness of a sheet of
graphene (carbon atoms)

Graphene is a two-dimensional form of carbon crystal that is just one atom or 0.3 nanometres thick, with just 0.1 nanometre between each atom. Formed of a lattice of hexagonally arranged carbon atoms, each of which forms four strong bonds with its neighbours, it is a material with remarkable properties. A single sheet of graphene is a million times thinner than a human hair, making it invisible to the human eye; a stack of 3 million sheets would be just 1 millimetre thick.

Graphene is the world's thinnest, strongest and most conductive material, yet it is stretchable and flexible. It is 200 times stronger than steel but six times lighter. It absorbs only 2 per cent of light that falls on it, making it almost perfectly transparent. It is impermeable, even to the smallest gas particles, such as helium atoms or hydrogen diatoms. According to mechanical engineering professor James Hone of Columbia University, 'It would take an elephant, balanced on a pencil, to break through a sheet of graphene the thickness of Saran Wrap [cling film].' According to the Nobel Prize committee, a square-metre sheet of graphene would weigh less than a cat's whisker (0.77 milligrams) but could support a 4-kilogram cat.

Scotch tape science

The existence of graphene was predicted by theory as early as 1947, when it was realised that graphite, a common form of carbon found in pencils, for example, is essentially a pile of graphene sheets. However, no one could work out how to get from graphite to graphene. It turns out that the answer is incredibly simple. In 2004, Andre Geim and Kostya Novoselov, Russian-born researchers at the University of Manchester, were discussing the technique used by scientists to polish graphite surfaces, using sticky Scotch tape to peel off the outer layers. They realised that the tape, which is usually discarded, actually held immense value – thin layers of graphene. Using repeated Scotch tape peels, they obtained flakes of single sheet graphene and published a landmark paper proving their success through electrical measurements. Six years later they shared the Nobel Prize in Physics for their breakthrough. Anyone can use the Scotch tape method to produce their own graphene, and in fact if you have ever drawn with a pencil you have almost certainly created graphene.

The combination of strength, lightness and conductivity has excited much speculation about applications of the new material, but research into other exotic forms of carbon – fullerene 'buckyballs' and carbon nanotubes – advises caution. Carbon nanotubes are essentially rolled-up sheets of graphene, which offer similarly revolutionary applications to graphene, but it has proved extremely difficult to produce quantities of nanotubes with uniform properties. Graphene may prove similarly difficult to process at sufficient scale for it to be useful in practice.

▼ The structure of graphene, showing the hexagonal lattice of carbon atoms, arranged entirely in a single plane, leading to graphene being dubbed the world's first two-dimensional material.

1.229

Density of the Earth's atmosphere (kg/m^3)

1.229 kilograms per cubic metre is the density of the Earth's atmosphere at SLS, which stands for 'sea level standard' or 'sea level static', depending on the source.

Stipulating the altitude at which the measurement is made is vital because density depends on temperature and pressure, and both of these vary with altitude. Normally the density (denoted by the Greek letter rho, ρ) of a gas is given at standard temperature and pressure (STP), which is understood to be atmosphere of pressure at 0°C, but this would be an atypically low temperature for the atmosphere at the Earth's surface, so sea level standard is preferred as the reference altitude. Because movement of air (i.e. wind) can also affect density, 'static' conditions may also be specified.

At STP, the density of dry air is 1.29 kg/m^3. The International Standard Atmosphere (ISA) is an idealised model of the atmosphere, used to allow fields such as rocketry and aviation to make uniform and comparable approximations. In the ISA model, the temperature at sea level is 15°C, so the density of air is 1.275 kg/m^3. As the temperature increases, density decreases, so at 20°C and 1 atmosphere, the density of dry air is 1.2041 kg/m^3. Counter-intuitively, humid air is less dense than dry air, because every additional water vapour molecule present displaces a heavier oxygen or nitrogen molecule. The atmosphere on Venus is 90 times denser than on Earth, while the Martian atmosphere has a density just 0.7 per cent that of Earth.

+2
Oxidation number of copper (II)

+2 is the oxidation number or state of copper when it donates/ loses two electrons in forming a chemical bond with another atom/ion. The oxidation number or state is the number of electrons gained or lost when an atom forms a covalent or ionic bond. It shows the role played by an atom in a redox reaction, where one reactant is reduced and the other is oxidised. Reduction means gaining one or more electrons and becoming more negatively charged; oxidation means losing one or more electrons and becoming more positively charged.

For instance, when copper (Cu) reacts with oxygen (O) to form CuO, known as cupric oxide, the copper atom donates two electrons to the oxygen atom. The copper is oxidised and the oxygen is reduced. Oxidation increases the oxidation number of the copper by two, while reduction reduces the oxidation number of the oxygen by two. In their elemental states, all atoms have an oxidation state of zero, so in the formation of cupric oxide the copper atom increases its state to +2. In the nomenclature of inorganic chemistry, the oxidation number of an element is indicated by a Roman numeral in parentheses. Thus, the technically correct name for this oxide of copper is copper (II) oxide.

Oxidation numbers are only given in this way for elements that may exist in more than one oxidation state, which is why it is important to give the state. Copper can also exist in an oxidation state of +1, where it donates just one electron, and so it can form a different oxide, Cu_2O, cuprous oxide, aka copper (I) oxide.

▲ A nugget of native copper – i.e. naturally occurring in its free, unoxidised form. This is rare among metals, and helped make copper one of the first to be worked by humans.

2

Superior air supplied by Joseph Priestley to a mouse (oz)

In 1774, Joseph Priestley 'procured a mouse, and put it into a glass vessel, containing two ounce-measures of the air' – the 'air' in question was oxygen, and Priestley was amazed to find that it sustained respiration for four times longer than 'common' (atmospheric) air.

Priestley was a dissenting minister whose interest in natural philosophy had been encouraged by Benjamin Franklin. A job next door to a brewery introduced him to pneumatic chemistry and he became famous after discovering the carbonisation of drinks, where carbon dioxide is forced into liquid under pressure so that it becomes effervescent. Thus he invented soda water 'a service to naturally, and still more to artificially, thirsty souls', according to Thomas Huxley, Victorian scientist and promoter of science.

A series of experiments on 'airs' contained within enclosed spaces (upturned glass bells), using instruments related to respiration and combustion, such as candles, mice, burning glasses (a means of focusing rays of sunlight to produce localised ignition) and green plants, put Priestley on the track of various components of the atmosphere. It was well known that a burning candle or breathing mouse soon exhausted the life/combustion-supporting properties of common air when confined to a closed vessel. In 1771, Priestley showed that these properties could be restored by a green plant, concluding that the atmosphere's ability to support life/combustion was being continually 'repaired by the vegetable creation'.

▲ Examples of the apparatus Jospeh Priestley used to capture and study gases.

To make the proof complete

In 1774, for reasons he himself described as vague – 'I cannot ... recollect what it was that I had in view in making this experiment; but I know I had no expectation of the real issue of it' – Priestley used his burning glass to heat mercury oxide placed in a closed vessel. To his surprise, he wrote, 'a candle burned in this air with a remarkably vigorous flame'. Although it was able to support especially bright combustion, he was initially 'perfectly satisfied of its being common air ... though, for the satisfaction of others, I wanted a mouse to make the proof quite complete'.

Accordingly, he shut up the mouse with two ounces of the new air and watched. 'Had it been common air, a full-grown mouse, as this was, would have lived in it about a quarter of an hour. In this air, however, my mouse lived a full hour ... and appeared not to have received any harm from the experiment.' It was not common air, but 'superior air' – Priestley had discovered oxygen. In fact he was not the first, as the Swedish chemist Carl W. Scheele had isolated what he called 'empyreal air' ('fire air') prior to 1771, but did not publish his findings until after Priestley. Also, it would be the great French chemist Lavoisier (see page 136) who would give the new air its modern name. Priestley subscribed to the phlogiston theory (see page 33), and believed he had discovered dephlogisticated air, a gas that could support combustion so well because, being devoid of phlogiston, it readily absorbed the stuff (phlogiston). Lavoisier showed how the new air was the key to a new theory of combustion, in which burning involved combining with a substance, not giving one off. Because he believed that this new element was present in all acids, he gave it a Greek name meaning 'acid-maker': oxygen.

▲ Simplified version of Priestley's 1771 experiment. At the top, both the burning candle and the respiring mouse quickly exhaust the available oxygen. At the bottom a green plant is able to restore the oxygen, keeping both candle and mouse alive.

3

Number of principles of Paracelsus: sulphur, mercury and salt

The *tria prima*, or three principles, are the basic elements of all bodies, in the metaphysical system of the 16th-century physician and alchemist Paracelsus.

Paracelsus was the Latin pseudonym of the Swiss-born Philippus Aureolus Theophrastus Bombastus von Hohenheim (1493–1541), a major figure in the evolution of the mystical art of alchemy into the science of chemistry. A celebrated but controversial physician, Paracelsus spent his career slaying sacred cows and upsetting people, but also laid some of the foundation stones for what would later become chemistry. Building on the theories of the Persian physician and alchemist Jabir ibn Hayyan (c.721–c.815), Paracelsus sought to update the four earthly elements of Aristotelian dogma (see page 54) to reflect more modern concepts about the nature of matter.

In the body of work associated with Jabir, Muslim alchemy had developed an Aristotelian theory about the origin of metals. Aristotle described the formation of metals in the Earth as arising from the combination of two 'exhalations' created by the action of sunlight. Jabir formalised these exhalations into two principles, called sulphur and mercury, but not to be literally understood as the physical substances of the same name. This 'dyad theory' suggested that the nature of a metal depended on the proportions and purity of the two principles, offering the hope that by purifying and remixing them, one type of metal could be transformed into another (such as turning base metals into gold).

▲ Key figures in the development of alchemy, the occult art of spiritual and material transformation, into chemistry, the science of matter, Jabir ibn Hayyan (below, known in the West as Geber) and Paracelsus (above) developed theories of fundamental principles that foreshadowed the elements of chemistry.

Added salt

To the dyad Paracelsus added a third principle, salt, to represent living bodies and the organic world, thus extending the scheme to cover all substances and not just metals. In the mystical and somewhat obscure model advanced by Paracelsus and his followers, every object comprises a 'low' salt principle, linked to solidity and dryness; a 'high' mercury principle, linked to fluidity, smokiness/volatility and transformation; and an intermediary sulphur principle, linked to oiliness and flammability, which connects the 'high' and the 'low'. Paracelsus overlaid these three principles onto the four Aristotelian elements to create twelve basic categories, extending their explanatory power to the human body and medicine. Subsequent alchemists added two more principles – phlegm and earth – making the triad into a pentad.

The extent to which this metaphysical model had much explanatory power or practical utility is doubtful. Its real importance was in allowing scholars to move beyond the intellectual straitjacket of Aristotelian dogma, opening up the space for new approaches to natural philosophy. Alchemists were starting to pay more attention to the phenomena they observed in their laboratories, with less concern for the pronouncements of classical authority. The proliferation of new elemental schemes and models was a consequence of this increased focus on experiments. Although the imminent birth of scientific chemistry in the work of men such as Robert Boyle (see page 134) would be in part a reaction against dyads, triads and pentads – 'the Chymical Doctrine as 'tis generally taught by the vulgar Chymists,' he wrote in his landmark book *The Sceptical Chymist* – it was also the direct result of the practices that gave rise to them, namely experimentation and applied chemistry in fields such as medicine and metallurgy.

▼ The three principles of the Paracelsian *tria prima* should not be confused with the literal materials after which they are named – mercury, sulphur and salt shown below.

4

Number of states or phases of matter

There are four states or phases of matter: solid, liquid, gas and plasma. The first three are familiar enough from everyday life, but the fourth was only recognised in the 19th century.

As the temperature of matter increases, it undergoes phase changes. At low temperatures, particles (atoms or molecules) are stationary enough to form bonds that lock them in place, forming a solid. Applying heat causes the particles to become more energetic and move about more, so that bonds between them weaken and can be constantly broken and re-formed. The solid changes phase to a liquid. Still more heat causes the particles to move about too fast to make any bonds, and they whizz about independently. Liquid has changed phase to gas. Applying still more heat – or other sources of energy, such as magnetic or electrical fields – causes electrons to jump or be stripped from their nuclei, changing the gas to a soup of charged particles (ions and electrons). This is the fourth phase of matter, known as plasma.

▲ William Crookes invented his eponymous tubes to investigate the relationship between electricity and partial vacuum, resulting in an instrument that could fire beams of electrons at gas molecules.

Worthy of Newton

The first three phases were recognised at least as early as the 18th century, when the discoveries of Scheele, Black, Cavendish and Priestley proved that there were many 'airs' or gases to set alongside solids and liquids. Even before this, the classical elements might be said to allude to phases of matter, with earth, water and air equating to solid, liquid and gas. Recognition of the fourth state first came with discoveries in electromagnetism by the great English scientist Michael Faraday (1791-1867). He described what he called 'radiant matter', writing: 'If now we conceive a change as far beyond vaporisation as that is above fluidity, and then take into account also the proportional increased extent of alteration as the changes rise, we shall perhaps, if we can form any conception at all, not fall far short of radiant matter... The simplicity of such a system is singularly beautiful, the idea grand, and worthy of Newton's approbation.'

The physical basis of the universe

Faraday notwithstanding, it is the English scientist Sir William Crookes (1832-1919) who is generally recognised as the man who 'discovered' the fourth state of matter. He was investigating the effect of electric currents on a partial vacuum, passing cathode rays (streams of electrons) through rarefied gas contained in glass tubes named after him, and found that the gas started to exhibit exciting properties, such as glowing and conducting electricity. Following Faraday, he described the contents of the Crookes tubes as a 'radiant matter', a 'Fourth state of Matter'. 'In studying this Fourth state of Matter,' he famously wrote in an article in the journal *Nature*, 'we seem at length to have within our grasp and obedient to our control the little indivisible particles which with good warrant are supposed to constitute the physical basis of the universe.'

In 1928, the Nobel-Prize-winning American chemist and physicist Irving Langmuir (1861-1957), dubbed this fourth state 'plasma', and it is now known that plasma is by far the most common state of matter in the visible universe. It makes up the stars and occupies much of the space between them, accounting for over 99 per cent of the visible universe. Plasma can be found everywhere from the exotic to everyday life, both natural and man-made: in lightning, the northern and southern lights, the solar wind, plasma TVs, neon signs and fluorescent lights, fusion reactors and rocket exhausts.

▼ The basic outlines of a toroidal (doughnut-shaped) fusion reactor, in which a plasma is confined by a magnetic field and heated until fusion occurs.

Poloidal field magnet

Toroidal field magnet

Vacuum chamber

Plasma

4

Valence of carbon

The valence or valency of carbon is 4. Valence is a measure of the combining power of an atom; the number shows how many bonds an atom can form.

Explaining and systematising the rules governing bonding between elements was one of the great challenges of 19th-century chemistry. With a greater understanding of the structure of the atom and the role of electrons in chemistry has come an understanding of valence. Valence is determined by the number of electrons in the outer electron shell of an atom (see pages 60-61 for more on electron shells), which is known as the valence shell. This number follows periodic rules, so elements in the same group of the periodic table have the same valency. Because of the octet rule (see pages 60-61), the highest valence an element can have is 4; this is the valence of elements in group 4, most notably carbon, which has six protons and therefore six electrons: two in its inner shell and four in its outer, valence shell.

The high valence of carbon underpins its remarkable properties, meaning that it is able to form four covalent bonds with other atoms, including other atoms of carbon. The bonds can be single, double or triple, and carbon atoms can join up to produce almost indefinitely long chains, which act as backbones to which other elements can attach. Carbon's limitless potential for making compounds gives rise to an entire sphere of chemistry – organic chemistry – and has made carbon the basis of all life on Earth.

4.186

Specific heat of water

(joule/g°C)

It takes 4.186 joules to heat up 1 gram of water by 1°C. This is the definition of a calorie, and is unusually high, with major consequences for the ecology of Earth.

The specific heat or specific heat capacity of a substance is the amount of energy required to raise a fixed mass by a fixed degree of temperature. For standardisation purposes, specific heat is given using a mass of 1 gram and a temperature change of 1°C. The calorie is defined as the amount of energy needed to raise the temperature of a gram of water by 1°C. The heat capacity of an object is its specific heat multiplied by its mass, and hence has units of joules/°C; for instance, the heat capacity of 1 litre (1,000 grams) of water is 1,000 × 4.186 = 4,186 joules/°C.

Water has an unusually high specific heat capacity; it is five times higher than aluminium and roughly ten times higher than iron or copper (it takes just 0.385 joules to heat a gram of copper by 1°C). The high heat capacity of water helps to regulate the temperature of bodies of water, so that while air temperatures vary widely across the Earth, sea temperatures vary much less, and a given body of water will undergo relatively small temperature changes whatever the weather. This is good for aquatic life, which has less extreme conditions to deal with than terrestrial life, and helps to regulate the climate of the Earth as a whole and of coastal areas in particular.

5

Number of elements in Aristotelian cosmology

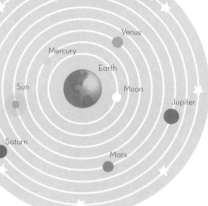

The classical philosophers of ancient Greece suffered from element inflation. In the first book of his *Metaphysics*, Aristotle himself traces the development of thinking on the question of the elements: 'That of which all things that are consist, the first from which they come to be, the last into which they are resolved (the substance remaining, but changing in its modifications).' Thales, Aristotle explains, was 'the founder of this type of philosophy; he says that the principle is water', on the basis that earth rests on water. Anaximenes and Diogenes make air the first principle, 'prior to water', while 'Hippasus of Metapontium and Heraclitus of Ephesus say this of fire'. Empedocles adds a fourth element, earth, and says that none has priority, 'for these, he says, always remain and do not come to be, except that they come to be more or fewer, being aggregated into one and segregated out of one'.

To the four elements of Empedocles, Aristotle added a fifth, aether or ether, found only in the heavenly sphere. In his cosmology, the spheres of Earth and the heavens are profoundly different; the former lies at the centre of the universe, while above and around it circle the crystalline spheres of the celestial bodies and the fixed stars, composed of perfect, unchanging aether. This cosmology would prove to be perhaps the most influential model in history, coming to be regarded as dogma in Western Europe and governing theorising about chemistry and the other sciences until the end of the Middle Ages in the late 15th century.

▲ Simplified diagram of the Aristotelian cosmology, with the Earth at the centre and the celestial bodies in concentric crystalline spheres, surrounded by the aetheric sphere of the fixed stars.

6

Period of the lanthanides

Period 6 of the periodic table includes the lanthanides, a family of rare earth metals. These were originally identified from an ore discovered at the mining village of Ytterby in Sweden in 1787. At this time, ores of the metals (aka 'earths') appeared to be hard to find, so they were known as the rare earths. In fact, rare earth metals are relatively abundant (the world reserve is estimated to be about 150 million metric tons, in the form of rare-earth oxides); they are much more common than platinum-group metals, for instance.

The lanthanides are a family of chemically similar silvery-white, mostly soft metals, comprising 15 elements ranging from lanthanum (La), atomic number 57, to lutetium (Lu), atomic number 71, and including some with widespread commercial and scientific use, such as cerium, neodymium and europium. The arrangement of electrons in their orbitals means that the outermost orbitals, which have the most effect on their chemistry, are full while the orbital that is further in is filled as the family progresses. This in turn means that all the lanthanides are chemically similar, even while they can vary widely in physical properties, particularly their properties regarding light and magnetism.

Their most common applications are as catalysts for oil refining, catalytic converters in cars and alloys for permanent magnets, but they can also emit sharply defined colours of visible light, so have been important in TV screens and optic fibres.

6,6

Designation of the polyamide used to create nylon

6,6 was the polyamide selected in 1935 as the basis for the creation of nylon, by DuPont and their lead researcher Wallace Carothers. The numbers refer to the numbers of carbon atoms in its precursors, hexamethylenediamine and adipic acid.

In the early 20th century, one of the great frontiers of chemistry was the mysterious and complex world of organic chemistry, and one of the great riddles of organic chemistry was the nature of long-chain molecules known as polymers. Polymers are large or macromolecules made from repeating subunits; they are common and essential in nature, from DNA to cellulose (see pages 163 and 159). In the early days of organic chemistry, it remained unclear how the subunits of a polymer are held together. Were they joined by the same types of chemical bond as much smaller compounds, building up into long chains, or were they peculiar colloidal systems held together by special forces, unique to the living world?

▲ As the first viable artificial silk, nylon caused a sensation in women's fashion with the popularity of nylon stockings, even leading to 'nylon riots' when demand outstripped supply in stores.

Purity Hall

Polymers based on existing natural ones had been produced since the 19th century, with forms of artificial silk based on cellulose. But to address the polymerisation question with a purely synthetic approach was the bright idea of brilliant and troubled American chemist Wallace Carothers. In 1927, Carothers was head-hunted from Harvard by the giant industrial chemical company DuPont to head up the organic chemistry side of their new 'blue sky' research division, nicknamed 'Purity Hall'. Writing to accept the job, he detailed his first research project: 'I have been hoping that it might be possible to tackle [the polymer] problem from the synthetic side. The idea would be to build up some very large molecules by simple and definite reactions in such a way that there could be no doubt about their structures.'

Carothers soon succeeded, reacting together alcohols and acids to produce esters and polymerising them to give short polyesters. His work, according to colleague Julian Hill, 'finally laid to rest the ghost ... that polymers were mysterious aggregates of small entities rather than true molecules'. But this landmark work was only the beginning. In 1930, Hill made long polyesters from which threads could be drawn out: the first purely synthetic fibres. They were unsuitable for commercial purposes because of their low melting point; it would be impossible to iron or launder them.

Nylon is born

Carothers reasoned that using amides in place of esters might produce commercially viable fibres, since amides have much higher melting points. The initial attempts were unsuccessful and it would be four years before Carothers hit upon a suitable precursor, during which time his fragile mental health had begun to deteriorate, with spells of severe depression, breakdowns and stays in clinics. In 1934 the first polyamide fibre was drawn: nylon had been created. But the precursor used was too expensive to be viable, so Carothers and his team assessed over a hundred alternatives, whittling it down to two options: polyamide 5,10, made from pentamethylene diamine and sebacic acid; and polyamide 6,6. The former was Carothers's preferred option but the latter was cheaper, and by 1938 polyamide 6,6 was being consumed in industrial quantities to feed the machines in a new factory capable of producing nearly 5,000 tonnes (about 11 million pounds) of nylon a year. Carothers had taken his own life the year before, swallowing cyanide from a phial he had carried about for years.

▼ DuPont's groundbreaking research at Purity Hall continued after the loss of Carothers; here a chemist watches an organic reaction at the Wilmington research facility in the 1950s.

6.941

Atomic mass of lithium (amu)

6.941 is the atomic mass of lithium, the lightest metal and the third smallest element. Lithium, atomic number 3, has just three protons in its nucleus, alongside four neutrons (although in nature just over 7.5 per cent of lithium occurs as the isotope lithium-6, which has only three neutrons). Along with hydrogen and helium, lithium is the only element to have been formed during the Big Bang, or at least the first three minutes of the universe's existence.

It is a silvery metal with a very low density – just half that of water – and is violently reactive, so that it is not found in its free state in nature. It was discovered in 1817, when Swede Johan August Arfwedson analysed a mineral found in Sweden 20 years earlier, which had been observed to impart an intense crimson colour to the flame when thrown on a fire. Arfwedson was able to deduce the existence of an alkali metal similar to sodium but lighter, but could not isolate it. Its name derives from the Greek for 'stone', because it was the first alkali metal to be found in mineral rather than plant sources.

Lithium has the highest heat capacity of any solid element, and so is used in heat transfer applications and alloys. It also has a very strong electrochemical potential, making it the most energy-dense option for batteries; lithium ion batteries are now essential to a host of technologies from mobile phones to electric cars. A lithium ion battery can store up to four times as much energy as a comparably sized nickel–cadmium battery, and deals better with temperature fluctuations and long-term storage.

▲ Petalite, a lithium bearing mineral that is an important source of the metal.

7

Atomic number of nitrogen

Nitrogen (atomic number 7) comprises the majority of the Earth's atmosphere and is vital to life due to its role in proteins and DNA. It accounts for about 2.5 per cent of all biomass on the planet and is the fourth most abundant element in the human body.

Nitrogen is the fifth most abundant element in the universe and makes up 78 per cent of the Earth's atmosphere, which contains around 4,000 trillion tonnes of nitrogen. As well as being highly unreactive and non-combustible, nitrogen is colourless, tasteless and odourless and does not dissolve well into liquids. It was discovered several times in the 1770s, mainly by a process of subtraction: atmospheric air was treated to remove the other main components – carbon dioxide and oxygen – and what was left was characterised variously as noxious, spent or burnt air. Lavoisier called it *azote*, from the Greek meaning 'without life', because it could not support organic life; but in 1790 it was named nitrogen by the French chemist Jean-Antoine-Claude Chaptal after it was identified as a constituent of nitre (potassium nitrate).

Triton, the largest moon of Neptune, also has a lot of nitrogen, but it is so cold that the nitrogen has frozen on the surface as rock-hard ice. The ice itself is transparent to what little sunlight falls on Triton, but dark patches caused by rocks and impurities cause local warming, which vaporises the nitrogen ice, releasing vast geysers that push ice and dust particles 8 kilometres above the surface.

8

Number of electrons
in the octet rule

The octet rule governs the most stable electron configuration for the representative elements (the elements that are not in the transition metal or inner-transition metal blocks), particularly the first 20 elements. It thus helps to predict what kinds of ions and compounds these elements will form.

Like water running downhill to find the lowest spot in the local landscape, thus minimising the potential energy it has as a result of gravity, matter seeks to achieve the lowest energy state possible given its 'local landscape' (the electron configurations available). This force to achieve the lowest, most stable energy state drives the basic principle of chemical bonding, which is the tendency for electrons to distribute themselves in space around atoms so as to lower the total energy of the group.

Seeking nobility

The elements in the s and p blocks of the periodic table have outer valence shells of electrons consisting of only the s and p orbits, which can contain two and six electrons respectively. Thus their valence shells can have up to eight electrons, and thanks to the quantum mechanical properties of electrons, a full valence shell is the most stable, lowest energy configuration. The noble or inert gases make up the furthest right column of the periodic table. They have full valence shells, each containing eight electrons, and this is why they are so inert (unreactive). The octet rule is an organising principle that helps explain how other s and p block elements form ions and bonds. It states that elements gain or lose electrons to attain an electron configuration of the nearest noble gas, with which they then are said to be isoelectronic.

For instance, the alkali metals of group 1 of the periodic table have a single electron in their outer, valence shells, while the next shell in will be complete, so the easiest way for them to obtain the electron configuration of the nearest noble gas is to give up the electron, leaving them with a complete shell now outermost. Sodium, for example, atomic number 11, will readily give up its outer electron, leaving it with the electron configuration of the noble gas neon, atomic number 10. Hence sodium readily forms ions with a charge of +1. Chlorine, with seven electrons in its valence shell, readily accepts an electron to give it the same electron configuration as argon, and so tends to form ions with a single negative charge. The octet rule makes it easy to predict that sodium and chlorine will form a compound with the formula NaCl, or more precisely the ionic formula Na^+Cl^-.

Lewis diagrams

The octet rule also applies to formation of covalent bonds. In a covalent bond, an electron is shared between the two bonded atoms. For instance, in the molecule carbon dioxide, the carbon atom with its valence shell containing four electrons (see page 52) is looking to gain a further four electrons to make it isoelectronic with neon, while oxygen needs two electrons to complete its octet. In CO_2, the carbon atom shares two electrons with each of the oxygen atoms, and each of them shares two with the carbon; thus each atom has at least a share of eight valence electrons. A Lewis electron dot diagram is a handy way to show the valence electrons in a covalent compound; each atom in the diagram should have eight electrons around it.

▼ A Lewis diagram of carbon dioxide, showing how the carbon atom forms two covalent bonds with each of two oxygen atoms, allowing each atom in the molecule to fulfil the octet rule with its valence (outer) electrons.

8.314

The ideal gas constant

(J/K mol)

8.314 joules per degree kelvin per mole is one way of defining the ideal gas constant, the constant that relates how, in a fixed quantity of gas, temperature is proportional to the product of pressure and volume. In units of J/K mol, the ideal gas constant, known as R in scientific notation, relates energy to temperature, and hence recurs in the mathematical descriptions of many laws and phenomena.

In the early days of chemistry as a science, quantitative analysis was much easier with gases than solids or liquids, thanks to the type and resolution of measuring technology available. Hence, many of the initial breakthroughs in establishing quantitative laws of chemistry came in the realm of gases.

The gas laws

The first of these great breakthroughs came in the work of Robert Boyle (see page 134), who measured the relationship between pressure and volume when air was compressed at a constant temperature. He found that there is an inverse relationship, such that the product of volume (V) and pressure (P) remains constant. Hence, if you squeeze some air into half its original volume, its pressure will double. This became known as Boyle's Law, aka the Pressure–Volume Law, and was the first of the gas laws.

Subsequently the French scientists Jacques Charles and Joseph Gay-Lussac (see page 155) showed how pressure and volume

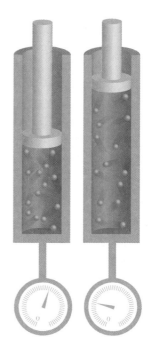

▼ Simple graphical explanation of Boyle's gas law, showing how, for a gas-filled piston, pressure (indicated by the gauge) varies inversely with volume (indicated by how far out the piston is pushed).

related to temperature. Charles' Law states that at constant pressure, volume is directly proportional to temperature (T), while Gay-Lussac's Law states that at constant volume, pressure is directly proportional to temperature. These and Boyle's Law apply to given amounts of gas. Italian lawyer and scientist Amedeo Avogadro (see page 77) provided the final piece of the puzzle, with his law showing how volume is directly proportional to the amount of gas (n): if you put twice as much gas into a bag, it will expand to twice its volume. The amount is measured in moles (mol), and is a measure of the actual number of atoms/molecules of gas present.

The combined gas law

Putting together the gas laws gives the equation PV/nT = R, where R is the constant that relates the other variables together. Because this 'combined gas law' refers to a so-called ideal gas, which obeys all the gas laws exactly, R is referred to as the ideal gas constant. It is also known as the universal gas constant, molar gas constant or simply, gas constant.

In this formulation, the ideal gas constant has units of (pressure × volume)/(amount × temperature), or atm L/mol K, and has the value 0.0821 atm L/mol K. The complete equation can be used to solve all manner of problems in the physics and chemistry of gas, including the likely formulae of gaseous compounds based on their weight, and of course working out the volume, pressure or temperature of a gas given the other variables.

The value of the gas constant depends on the units. Because the product of pressure and volume is energy, R can be given in units of energy/(amount × temperature), or joules/mol K. This in turn allows R to relate temperature to kinetic energy. In the equation PV = nRT, RT is the kinetic energy of an average molecule, so that nRT is the total kinetic energy of all the gas molecules put together.

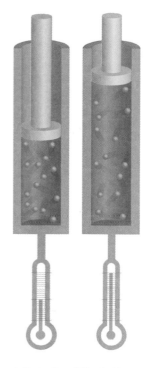

▲ Illustration of Charles' Law, showing how as the volume of a gas increases (as indicated by the height of the cylinder), so does the temperature, assuming that the pressure remains constant.

10

Number of points on Mohs scale of mineral hardness

Mohs is a scale devised by German mineralogist Friedrich Mohs in 1812 to grade the hardness of minerals. It ranks minerals on a scale of 1 to 10, by determining which minerals can scratch another mineral, with each point on the scale assigned to a mineral of known hardness. Other scales use more quantitative, less relative measures of hardness.

The Mohs scale gives minerals, particularly gemstones, a hardness ranking in Mohs. The ten points on the scale, from softest to hardest, are: talc, gypsum, calcite, fluorite, apatite, orthoclase (aka feldspar), quartz, topaz and beryl (which includes emerald), corundum (which includes ruby and sapphire) and diamond. A mineral can be scratched by another mineral of equal or greater Mohs hardness, and this is how Mohs grades are assigned. For instance, if a mineral can be scratched by orthoclase but not by apatite, its Mohs hardness is between 5 and 6. For comparison, the Mohs hardness of a steel file is around 7, while glass is around 6 and a small coin about 3. Your fingernail has a Mohs rating of around 2.5. Precious metals such as gold and silver have Mohs grades of around 3-4.

The Mohs scale has some drawbacks. It only works for materials that can sustain clean scratches: those that are fine-grained or crumbly will not give clean readings. More importantly, the scale is relative to the arbitrarily assigned mineral referents, and not every unit is equal. For instance, calcite and fluorite (3 and 4 on the Mohs scale) differ in hardness by only around

▲ Diamond gets its name from the same Greek root as 'adamant', meaning unbreakable. The extreme strength and hardness of diamond make it the natural choice for hardness testing equipment.

25 per cent, whereas the difference in hardness between corundum and diamond (9 and 10 on the scale) is over 300 per cent, even though it is just one Mohs unit.

Under pressure

More quantitative scales of hardness use indentation, where a hard point, such as a diamond, is pressed into the test material and the load required to produce a certain depth of indentation is measured. The Vickers hardness test, devised at the Vickers Company in 1921, uses a pyramid-shaped diamond point to create indentations. Measurements of force applied and the size of the indentation are combined to give a Vickers Pyramid Number (HV) or a Diamond Pyramid Hardness (DPH) rating.

Hardness records

Another measure of hardness is compression strength under indentation, measured in pascals (units of pressure, Pa). A 2009 study in *Physical Review Letters* set new records for hardness by comparing diamond to two crystals that change their structure under compression loading, so that they become stronger as they are crushed. Diamond achieves indentation strength of 97 gigapascals (GPa), while a material called wurtzite boron nitride reaches an indentation strength of 114 GPa, and one called lonsdaleite, aka hexagonal diamond (which, like diamond, is a crystal of carbon), yields an indentation strength of 152 GPa, which is 57 per cent higher than the corresponding value of diamond.

▲ Talc, also known as soapstone for its greasy feel, is the softest mineral - it can be scraped with a fingernail - and hence the reference for number 1 on the Moh scale of hardness.

11

Atomic number of sodium

11 is the atomic number of sodium, an alkali metal of group 1 of the periodic table, and one of the most important elements for animals, although it occurs in nature only in compounds, being too reactive to exist as the free element.

Sodium is a soft, waxy, silvery metal. A freshly cut surface acquires a patina of oxide within minutes, and sodium metal put into water reacts violently, liberating hydrogen from the water. Its high level of reactivity is due to its single valence electron, which it readily loses/donates in seeking to follow the octet rule (see page 60). This makes sodium a powerful reducing agent, with a reduction potential of -2.71 volts, placing it fifth in the list of the reduction potentials of the elements; lithium is top with a reduction potential of 3.04 volts.

▼ Sea salt is farmed by evaporating seawater to leave behind crystals of salt, which are mainly sodium chloride but will also have traces of magnesium, potassium and other elements.

Sea salt

The extreme reactivity of sodium explains why it is not found in nature in its metallic elemental state, but it is plentiful, accounting for 2.6 per cent of the Earth's crust, making it the sixth most abundant element on Earth. Most of this is in the form of sodium chloride (NaCl, which we know as table salt), and most of this is dissolved in the oceans, having been leached out of the crust over billions of years. If all the sodium chloride in the oceans were extracted and dried, it would cover the entire land surface of the Earth to a depth of 150 metres.

Another common compound of the element is sodium carbonate (Na_2CO_3, aka washing soda or soda ash), and it was this that was electrolysed by Humphry Davy in 1807 when he became the first to isolate the metal (see page 123). It appeared, he noted, 'to have the lustre of silver ... is exceedingly malleable and is much softer than any of the common metallic substances...'. Because it had been isolated from soda, Davy named it sodium, although he also toyed with the idea of calling it sodagen, following Lavoisier's naming scheme.

Mummy maker

Since ancient times, soda had been obtained from Egypt's Wadi El Natrun, or Natron Valley. Natron salt, as it was thus known, was a vital ingredient in the mummification process practised for millennia by Egyptian morticians. Natron is very strongly hygroscopic (it absorbs and holds water), and it was used to extract water from the body as part of the preservation process. The liver, lungs, stomach and intestines would be cut out of the body (the heart was left in place: as the seat of emotion and intelligence, it would be needed by the deceased as he attempted to navigate the afterlife), washed and packed in natron, and stored in canopic jars. Then more natron was packed into the body cavities and around the corpse. From this ancient name for soda, Berzelius derived the Latinate name for sodium, *natrium*, from which in turn derives its annotation, Na.

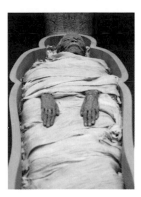

▲ Natron salt from Wadi El Natrun was an essential part of the mortuary technician's art in ancient Egypt; dehydrating the flesh halts the decaying action of microbes.

12.011

Atomic mass of carbon

12.011 is the atomic mass of carbon, but given that carbon provides the reference element by which the atomic mass unit is defined (see page 20), why does carbon not have an atomic mass of precisely 12? The answer is that carbon, like most of the naturally occurring elements, has more than one isotope.

'Isotope' derives from Greek roots meaning 'having the same place'. The term was coined in 1913 by British chemist Frederick Soddy, on the suggestion of Margaret Todd, to indicate atoms that occupy the same place in the periodic table but have different atomic mass and thus different physical properties. Isotopes of an element have the same number of protons in their nuclei, and thus the same atomic number and the same number of electrons, which in turn means they have the same chemical properties. But because they differ in the number of neutrons in the nucleus, they differ in mass.

In nature, carbon occurs primarily as the isotope carbon-12, which has six neutrons alongside the six protons, and it is this specific isotope that is used as the reference for atomic mass units. However, 1.07 per cent of naturally occurring carbon atoms have seven neutrons, making them carbon-13 isotopes. Because neutrons have atomic masses very slightly more than 1, the atomic mass of C-13 is 13.003. The atomic mass of naturally occurring carbon is thus the product of the proportions of these isotopes: $(12 \times 0.9893) + (13.003 \times 0.0107) = 11.8716 + 0.1391321 \approx 12.011$.

The atomic weight conundrum

The discovery of isotopes solved a long-standing puzzle in chemistry. In 1815, William Prout had noted that the atomic weights of the elements then measured all seemed to be whole numbers, with the atomic weight of hydrogen as the only common divisor. Accordingly, he suggested that all the other elements are aggregates of hydrogen atoms, which constitute the primordial unit of matter: the protyle. Prout's protyle was a key influence on Rutherford, leading to the conception of the proton, partially vindicating Prout's hypothesis, but in the 19th century more precise measurements of atomic weights had already shattered it. It became apparent that the atomic weights of the elements were not integers, but varied very slightly from whole numbers. In some cases they varied greatly: when chlorine was shown to have an atomic weight of 35.5, Prout's theory seemed dead in the water.

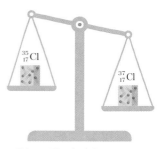

Chlorine-35 is the lighter isotope

The lighter isotope is deflected more in an electric field

The isotope solution

By the end of the 19th century, atomic doctrine suggested that all atoms of an element are identical, but the puzzle of the fractional atomic weights remained. Discoveries in radioactivity led to a solution, when substances were isolated with chemical properties identical to thorium but with different physical properties. Analysis showed that they had different atomic masses, leading Soddy to coin the term 'isotopes' (work that helped him win the 1921 Nobel Prize in Chemistry). Initially, isotopy was believed to be a property only of unstable radioactive elements, but it soon became clear that it applies to stable naturally occurring ones. In 1919, Francis W. Aston used his newly invented mass spectrograph to isolate two isotopes of neon and determine their atomic mass, and soon it was shown that chlorine also had two isotopes, finally solving the puzzle of its fractional atomic weight. Aston's spectrograph proved definitively that many elements exist as isotopes, and in 1922 he won the Nobel Prize in Chemistry 'for his discovery, by means of his mass spectrograph, of isotopes, in a large number of non-radioactive elements'.

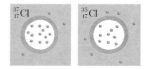

The lighter isotope more readily diffuses through a porous membrane

▲ Isotopes of the same element have identical chemical properties, but, as this graphic illustrates, their fractionally different atomic masses mean they differ slightly in their physical properties.

13

Approximate concentration of alcohol at which wine yeast inhibits its own fermentation (%)

13 per cent is approximately the maximum concentration of alcohol in wine, because concentrations above this inhibit the growth of the yeast that is fermenting the alcohol.

Fermentation is one of the oldest applications of chemistry in human history. Our prehistoric ancestors probably took advantage of naturally occurring fermentation to achieve intoxication, much as some animals do now. Early hunter-gatherers might have encountered naturally fermented fruits, and would almost certainly have inadvertently produced mead when diluting wild honey. The microbe responsible for this fermentation – yeast – became one of the first organisms domesticated by humans when they started deliberately to brew alcoholic beverages, although the yeast was not identified until the work of Louis Pasteur in the mid-19th century.

Yeast ferment sugars into alcohol when they are forced to metabolise in anaerobic conditions. In aerobic conditions they metabolise sugars much as animals do, to produce water and carbon dioxide, but in the absence of oxygen they have other metabolic pathways open to them and can convert glucose sugar ($C_6H_{12}O_6$) to ethyl alcohol (CH_3CH_2OH, aka ethanol) and carbon dioxide gas (CO_2). Alcohol, however, is mildly toxic, so above certain concentrations it inhibits the actions of enzymes and thus inhibits its own production. The precise concentration depends on the strain of yeast. For brewers' yeast, used to ferment beer, it is around 5-6 per cent. For wine yeast it is 12-14 per cent.

13

Age of Lavoisier's wife, Marie-Anne Pierrette, née Paulze, when he married her

In 1771, a young man intent on a career in science married the daughter of a colleague. He was Antoine Lavoisier and she was Marie-Anne Pierrette Paulze; she was only 13. Marie-Anne would go on to become the 'first lady of chemistry', though the true role she played in her husband's researches will never be fully known.

Marie-Anne's father, Jacques Paulze, was a tax farmer (private tax collector) in pre-revolutionary France. When his wife died, his daughter left convent school to serve as a hostess in his home. Here she attracted suitors, and her father encouraged the attentions of Lavoisier, a young man who had recently bought into the tax farming enterprise, and had been a visitor to their home. Paulze presented to the newly-weds a gift of scientific equipment, but Marie-Anne would prove to be Lavoisier's greatest help. She trained in chemistry, perfected her skill with languages so she could translate scientific papers, and honed her artistic talents by training in draughtsmanship. The second volume of Lavoisier's magnum opus, *Elements of Chemistry*, includes 13 plates detailing 170 pieces of laboratory equipment finely drawn to scale by Mme Lavoisier. It is also known that she recorded – and possibly managed – his experimental programme.

In 1794, Marie-Anne's husband and father were executed by the Revolution for their history of tax collecting, and she herself spent 65 days in prison. Undaunted, she reclaimed Lavoisier's papers on her release and eventually oversaw their publication.

14

Number of electrons
in the f orbitals

There are seven f orbitals, each capable of holding two electrons, so that the f orbitals can hold 14 electrons.

The classic image of the atom is the 'solar system' model that shows electrons orbiting the nucleus like planets around the Sun. However, electrons do not occupy orbits around the nucleus, they occupy orbitals, which are fundamentally different. An orbit is a defined path, usually circular or elliptical. The principles of quantum physics, such as the Heisenberg uncertainty principle, show that it is impossible to know both the position and momentum of an electron (i.e. where it is *and* where it is going next). All that can be determined is the probability that it will be somewhere. The zone of high probability of electron position is known as an orbital: a region of space around a nucleus where an electron of a given energy state is most likely to be found.

Electrons can have different energy levels, each with different potential orbitals. The first four energy levels have 16 possible orbitals between them.

▼ Illustration showing the strange and complex shapes of the electron orbitals at each energy level.

17

Speed of light travelling through a BoseEinstein condensate (m/s)

Although there is an upper limit on the speed of light, there is no lower limit. Experiments with an exotic form of matter called a Bose-Einstein condensate have shown that light can be slowed or even stopped altogether.

Work by Albert Einstein and the Indian physicist Satyendra Nath Bose led, in 1924, to the proposal of a new kind of matter. They theorised that atoms cooled to within a few billionths of a degree of absolute zero will see their quantum states lock in step, so that they can all be described by the same waveform, becoming a single super-atom: a Bose-Einstein condensate (BEC). Not until the 1990s was it possible to hit low enough temperatures to create a BEC.

In 1999, Lene Hau and colleagues at the Rowland Institute for Science in Cambridge, Massachusetts, used a vacuum with a pressure hundreds of trillions of times lower than that of air at Earth's surface, and temperatures of just a few nanokelvins (see page 28), to slow sodium atoms until they formed a BEC. Using interfering lasers, they then succeeded in slowing a light beam 20 millionfold from nearly 300 million m/s to just 17 m/s, or as Harvard University put it, 'the speed of a minivan in rush-hour traffic – 38 miles an hour'. Two years later, Hau's team was able to stop the light altogether, before releasing it with another laser pulse. It is believed that such feats could have practical applications for optical telecommunications, optical data storage and, if it were possible to fit BEC-creating technology onto a computer chip, quantum computing.

19

Number of mononuclidic elements

There are 19 mononuclidic elements, which is to say, 19 elements that are found naturally on Earth with only one stable nuclide (i.e. isotope). Such elements are important in the science of measurement, because their atomic masses are the same as their relative atomic masses (i.e. the mass of the element weighted to account for the abundance of naturally occurring isotopes).

Most of the elements occur in nature as a mixture of two or more isotopes: atoms of the element with the same number of protons but different numbers of neutrons (see page 69). This long confounded attempts to pin down their atomic masses with precision, since any naturally acquired sample of an element would include atoms that differed by one or more atomic mass units. Some elements, however, have only one naturally occurring isotope, allowing their atomic masses to be calculated with extreme precision.

The 19 elements are: beryllium, fluorine, sodium, aluminium, phosphorus, scandium, manganese, cobalt, arsenic, yttrium, niobium, rhodium, iodine, caesium, praseodymium, terbium, holmium, thulium and gold. Until 2003, there were 20 elements considered to be part of this list, but it was discovered that although bismuth is only found in nature as a single isotope, that isotope is in fact very minutely radioactive, which is to say, it has an unstable nuclide (with a half-life of 10^{19} years, about a billion times the age of the universe).

22
Atomic number of titanium

Titanium, element number 22 in the periodic table, is a silvery-white metal with a host of applications. It is as strong as steel but 45 per cent lighter; it is twice as strong and only 60 per cent heavier than aluminium, the other metal prized for combining strength and lightness. In its pure metallic form it is highly resistant to corrosion; even after 4,000 years of exposure to seawater, corrosion would penetrate no further than the thickness of a sheet of paper. Accordingly, it is widely used in maritime and aviation applications, and would be more widely used as cladding – as in Frank Gehry's Guggenheim Museum in Bilbao, Spain – were it not so expensive. Titanium can be given a startlingly colourful patina simply by application of an electric current and a weak acid, which forces the normally unreactive metal to form a thin layer of titanium oxide.

In powdered form, titanium oxide (titania) is a white pigment, so widely used (in everything from paint to sunscreen) that it accounts for 95 per cent of the end use of titanium production. As a patina, titanium oxide is largely transparent to light; it forms only in very thin layers, equivalent to just a few wavelengths of visible light. Light waves penetrate the oxide and bounce off the metal beneath, and they then interfere with the light that does reflect off the surface of the oxide. The interference produces strong colours, varying from yellow to blue depending on the precise thickness of the patina.

22.4

Molar volume for any gas (l)

At standard temperature and pressure (STP), the molar volume, VM, for any ideal gas is 22.4 litres. This means that a mole of any gas occupies the same volume.

The discovery of molar volume was one of the most important in the history of chemistry, revolutionising scientists' ability to quantify the chemical phenomena they were observing and laying the basis for atomic theory and accurate knowledge of formulae.

The riddle of the weights

At the start of the 19th century, chemistry seemed to have made tremendous strides to becoming a fully fledged science, complete with laws of nature underwritten by mathematical principles. Lavoisier destroyed the phlogiston doctrine and delineated the elements, while John Dalton's work on atomic theory had pointed the way to the precise formulation of compounds and, above all, the relative weights of the elements. The basis for this work was gravimetric analysis – i.e. weighing – which had revealed that, for instance, the proportions by weight of oxygen, sulphur and nitrogen combined with one unit of hydrogen are, respectively, 8, 16 and 5. Working on the assumption that nature always followed the simplest rules, Dalton assumed that compounds are formed of combinations of one atom of each element, so that, for instance, water must have the formula HO, and the atomic weight of oxygen must be 8.

▲ Amedeo Avogadro was the son of a Piedmontese nobleman who practised as a lawyer but taught himself science, later becoming a professor at Turin.

But what if the formula of water were actually H_2O, in which case the atomic weight of oxygen would be 16? With gravimetric analysis there was no way to know which was the right answer, but until the question was settled chemistry would struggle to progress further, since the true formulae of compounds could not be determined.

Avogadro's hypothesis

In 1811, Italian lawyer Amedeo Avogadro published an article in the *Journal de Physique*. In it he presented the solution to the riddle of how to determine true formulae and hence true atomic weights. His answer was to use volumetric analysis instead of gravimetric. Avogadro had read about Joseph Gay-Lussac's law of combining volumes, published two years earlier, which stated that gases combine to give gaseous products in simple integer ratios by volume. For example, two volumes of hydrogen gas combine with one volume of oxygen gas to give two volumes of steam. Avogadro realised that this finding could be explained if equal volumes of all gases, under the same conditions of temperature and pressure, contain equal numbers of molecules (in the sense of a particle that is either an atom or a molecule in the modern sense); this is now known as Avogadro's hypothesis. If the hypothesis is true, then one molecule of water must be formed from half a molecule of oxygen, and since, following Dalton, an atom cannot be split in half, the molecule must consist of two atoms of oxygen. In other words, oxygen gas is diatomic, the true formula of water is H_2O, the atomic weight of oxygen is 16, not 8, and the puzzle that had stymied all of chemistry was solved. Later it was determined that at STP, the volume of gas occupied by precisely one mole of molecules is 22.415 litres, but in fact Avogadro's revolutionary hypothesis was not appreciated for another 50 years.

Burette   Pipette

Beaker

Conical flask

Graduated volumetric flask

▲ Volumetric analysis, which makes use of equipment like that shown, helped overcome the limitations of gravimetric analysis.

22.59

Density of osmium (g/cm³)

With a density of 22.59 grams per cubic centimetre, osmium is the densest element that occurs naturally on Earth.

Osmium was named by its discoverer, English chemist Smithson Tennant, in 1803. He chose the name from the Greek root *osme*, meaning 'smell', because the newly isolated metal appeared to give off a foul odour. In fact, the metal in its powdered form quickly oxidises to give osmium tetroxide (OsO_4), which is responsible for the smell, is highly toxic and will damage tissue with which it comes into contact.

Osmium is notable for its extremely high density – it is twice as dense as lead, for instance. This helps to make it hard-wearing and resistant to frictional corrosion; its most common use is as an alloy to create small parts that need to be resistant to friction, such as gramophone-player needles, pen nibs and electrical contacts.

Density is an important concept in chemistry and the other natural sciences. It is a measure of how much mass is present in a given volume. For a regular-shaped object, density is easy to calculate by measuring the dimensions of the object to give the volume, and then weighing it. But for irregular-shaped objects, this calculation is not possible. Archimedes famously found an alternative method in the 3rd century BC, realising that a solid object would displace its own volume in water, making it easy to then calculate the density.

28

Atomic number of nickel

Nickel is element number 28 in the periodic table; its name comes from the German word for an imp or goblin, and references miners' beliefs about a nickel ore, which in turn reveals the roots of the science of chemistry in the world of artisans, miners and other non academic occupations.

Nickel is more abundant inside the Earth than in the crust. A lot of this 'internal' nickel probably arrived when huge asteroids were colliding to form the primordial Earth. Today, nickel is often found as a major component of metallic meteorites. In fact, most of the nickel mined in the world today, from a huge deposit in the Sudbury region of Ontario, Canada, may well be the result of an ancient meteor impact. The nickel content helps to keep metallic meteorites shiny and lustrous, which may account for the importance of such meteorites as objects of veneration in prehistoric times.

In Saxony, a spirit known as a nickel, a kind of gremlin or 'scamp', was blamed for the occurrence of a false copper ore accordingly known as kupfernickel ('copper scamp'). In 1751, Swedish chemist Axel Cronstedt analysed kupfernickel and isolated a hard white metal, the colour of which proved it was not copper; he named the new metallic element after its rascal forebear. Similarly, cobalt gets its name from the imps (Kobolde in German) held responsible for another false copper ore. From the days of Agricola and Paracelsus, Renaissance alchemists and pioneers of chemistry, mining folklore was an essential source of

▼ Nodules of nickel produced by electrolysis, next to a cube of the pure metal.

42.86

Earth's crust made up of silicon dioxide (%)

42.86 per cent of the Earth's crust is made up of silicon dioxide, aka silica, the most abundant compound on Earth. It is a ubiquitous component of rock and sand.

Silicon dioxide is a compound with a giant covalent structure (the same sort of structure that diamond has), which is very hard and has a very high melting point. It is also an excellent insulator. It forms crystals that can be very large, in which case they are known as quartz, a mineral that is generally milky and transparent but can take a huge variety of forms and colours. For instance, amethyst, rock crystal, agate, carnelian, flint, jasper and onyx are all forms of quartz.

Silicon dioxide is most familiar as small grains known as sand. In fact, sand that you find on the beach is not pure silica, as there will be grains of other minerals, which is why its colour varies. Estimating the number of grains of sand in the world dates back at least as far as Archimedes. In his work *The Sand Reckoner*, written partly to show off his mathematical skill, he invented new numbers to describe the number of grains of sand it would take to fill the cosmos, arriving at a number between 10^{51} and 10^{63}. On Earth alone there are an estimated 10^{24} (a trillion trillion) grains of sand. If all the silicon dioxide in the Earth's crust were extracted and ground into sand there would be a heap weighing 1.187×10^{22} kg, roughly the mass of Pluto.

45

Energy density of gasoline
(MJ/kg)

45 MJ/kg is one of the most important numbers in human civilisation: it is the energy density of gasoline (aka petrol), in megajoules per kilogram.

Gasoline is an essential source of energy for modern civilisation, mainly for powering road vehicles, but also for other engines and motors, from boats and tractors to home generators and aeroplanes, as well as domestic and commercial heating. Global oil consumption in 2014 was running at around 9.4 million barrels per day, or over 33 billion barrels a year, and continues to rise; not all of this is gasoline, but in terms of day to day impact on most people, gasoline is the most important petroleum product. Why is gasoline so important and ubiquitous? The answer is 45 MJ/kg. The extremely high energy density of gasoline makes it extremely useful as a source of energy; it is easy to carry around enough of it to power an engine for long periods, without the bulk of that engine's output going towards shifting the weight of fuel.

For comparison, an alkaline AA battery weighs around 25 grams and supplies roughly 10 kJ of energy; the same weight of gasoline contains roughly 1,125 kJ, or over a hundred times as much energy. Gasoline has an energy density around 50 per cent higher than coal, but with much cleaner burning and easier handling. In fact, gasoline is quite hard to set alight; it is gasoline vapour that is highly inflammable. This means that, if handled properly, gasoline is relatively safe to transport and distribute, despite its high energy density.

50-50

Mix of nitrous oxide and
oxygen in 'gas and air' (%)

A mixture of 50 per cent nitrous oxide (aka laughing gas) and 50 per cent oxygen, known as Entonox® or gas and air, is a commonly administered analgesic (pain relief agent) and sedative, often used in childbirth and ambulance settings. It is the modern sequel to one of the most important and entertaining chapters in chemical history.

Nitrous oxide (N_2O) is a colourless, odourless gas at room temperature. The precise mechanism of its effects when inhaled are not clear, but its analgesic effects are believed to be mediated by similar pathways to opiates – i.e. by triggering release of endorphins or mimicking their action – although it is not generally addictive. Usually, patients administer it to themselves by holding a mask or tube to their face or mouth, so overdose is impossible because loss of consciousness, which leads to dropping the applicator and cutting off supply, precedes asphyxiation. Even when used for anaesthesia, if properly administered nitrous oxide is safe, with rapid recovery and no after effects.

Exploring the airs

Many of these advantages were discovered by the great English chemist Humphry Davy at the beginning of his career, in an episode that would be formative in the developing character of the science of chemistry. Davy was a bright young man from the far-flung country of Cornwall, deep in the south-west of

England, but his precocious intellect, enthusiasm and ambition had caught the eye of figures in the scientific establishment and he was recommended for the post of Medical Superintendent at the Pneumatic Institute in Bristol. The Institute was a new project by the philanthropic and somewhat eccentric physician Thomas Beddoes, who was keen to apply the presumed medical benefits of recently discovered 'airs' to the needs of the unserved poor. Part of Davy's job was to undertake a proper clinical evaluation of the various airs, in the course of which researches he habitually experimented on himself.

Capable of destroying pain

The most promising of the new gases turned out to be nitrous oxide, discovered in 1772 by Joseph Priestley, who called it 'diminished nitrous air'. Davy inhaled ever larger doses of the gas. Its euphoric effects had him laughing and stamping his feet, and led to its common name 'laughing gas'. Larger doses caused intense 'trips' and eventually a brief spell of unconsciousness, from which he quickly recovered with no ill effect. He also noted the analgesic effects, recording in his laboratory notebook that 'sensible pain is not perceived after the powerful action of nitrous oxide' and using it to treat his own dental pain. In his 1800 paper, 'Researches chemical and philosophical; chiefly concerning nitrous oxide', he wrote: 'As nitrous oxide ... appears capable of destroying physical pain, it may probably be used with advantage during surgical operations.'

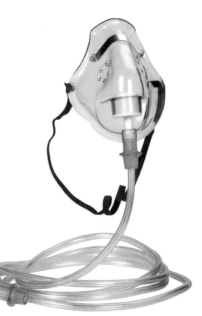

▼ Apparatus for administering gas and air; women in labour can hold the mask themselves and thus control their dose according to need.

Yet despite being on the verge of the discovery of medical analgesia and anaesthesia, Davy never made the final leap to a practical application. Losing interest in gases and the Pneumatic Institute, he moved to London to pursue electrochemistry (see page 122). It would be a further 40 years before American dentist Horace Wells made the first demonstration of dental treatment under anaesthesia, and his work eventually led to the widespread adoption of surgical anaesthesia, although ether was preferred to nitrous oxide because of the latter's risk of oxygen starvation with improper administration.

56

Atomic number of barium

Barium is element number 56 in the periodic table. It is extremely reactive, so is never found in nature as the free metal. Its sulphur compounds have intriguing phosphorescent properties, which made them magical curios in early 17th-century Europe.

Barium is found most commonly in the ore barite (barium sulphate, $BaSO_4$), the high density of which led to its name, derived from the Greek for 'heavy'. Around 1600, an Italian alchemist, Vincenzo Casciarolo of Bologna, began researches into samples of this curious ore, having learnt of their phosphorescent properties. Heating chunks of the ore with charcoal during the day produced a material that phosphoresced at night, glowing red for up to six hours. Casciarolo thought he had stumbled upon the philosopher's stone, a mythical wonder-working catalyst with the power to transmute base metals to gold. He was disappointed, but the barite pebbles became known as Bologna stones.

Barium was identified in 1774 as a new element by Swedish chemist Carl W. Scheele, who described 'an earth differing from all earths hithero known'. Scheele was unable to isolate it, partly because barium sulphate is insoluble in water. Davy first isolated the pure metal by electrolysing molten barium salts in 1808. The insolubility of barium sulphate, married to the fact that it is opaque to X-rays, makes it extremely useful as a diagnostic aid for radiography. Administration of a barium meal or enema makes it possible to clearly visualise the digestive system, after which the barium sulphate is completely expelled from the body.

60

Number of tubs of urine processed by Hennig Brand to discover phosphorus

In 1669, German alchemist Hennig Brand processed 60 tubs of putrefying urine to recover a tiny amount of white phosphorus: the first element with a known discoverer. With its unearthly glow, it caused a sensation.

The roots of modern chemistry lie in medieval and Renaissance alchemy, and this is particularly evident in the discovery of phosphorus. Phosphorus is a highly reactive metal that exists in several forms, or allotropes, including white, red and black. White phosphorus gives off a dim green glow, which gave rise to the technical term phosphorescence, meaning a substance that absorbs light and continues to re-emit it at different wavelengths after absorption (i.e. it glows after being illuminated). Ironically, this is not the actual mechanism by which white phosphorus glows; it is chemiluminescent, producing light through a chemical reaction, as vaporous phosphorus rising from the solid reacts to form an oxide.

Hidden gold

Brand had, like many alchemists before him, been searching for the philosopher's stone. He was a disciple of Paracelsus, who had promulgated various pseudo-scientific, mystical doctrines of the alchemical tradition. Among these was the doctrine of signatures, a belief that occult properties were revealed through 'signatures' in the natural world: associations and symbols discernible to the

▲ Joseph Wright of Derby's famous imagining of Brand's moment of illumination, titled *The Alchymist, in Search of the Philosopher's Stone, Discovers Phosphorus.*

initiated, such as the golden colour of some minerals and plants revealing an affinity with the precious metal gold. At the same time, there was a rich tradition of the alchemical potency of bodily fluids, including excrement (this has a strong basis in fact, given that excrement often contains high concentrations of chemically active substances such as nitrogen compounds). Accordingly, Brand believed that urine, with its yellow colour and chemical potency, might hold the key to the creation of the philosopher's stone.

In one of the most unpleasant experiments in chemical history, Brand acquired 60 tubs of urine and let them sit in a cellar until they 'putrefied', and then boiled the noisome contents down to a paste, producing what alchemists called microcosmic salt (ammonium sodium hydrogen phosphate). Vapours were drawn off through water to be condensed as the paste was heated with sand (producing sodium phosphite) and then charcoal (combining the phosphite with carbon to give sodium pyrophosphate and a residue of white phosphorus).

Match light

Brand named the waxy, glowing substance 'phosphorus', from the Greek for 'bringer of light'. Exhibiting his discovery brought him a living and he sold the secret of its production for high prices. Robert Boyle managed to deduce the process for himself, making his own white phosphorus, which he called *noctiluca*, 'cold light'. In the 1770s, Scheele showed that phosphorus could also be prepared from ground-up bone. Boyle had shown that white phosphorus, which is so reactive it can spontaneously combust in air and must be stored under oil, could be used to make matches, 150 years before they were properly invented. Red phosphorus, a more stable form of the element, is preferred in modern safety matches, where it is found in the strip against which the match is struck.

61

Number of elementary subatomic particles in the Subatomic Zoo

In the Standard Model of physics, there are 61 elementary particles: indivisible subatomic particles, some of which combine to make up other subatomic particles.

Their discovery began with the electron in 1897, followed by the proton and then, in 1932, the neutron. This was just the beginning of an avalanche. In the 1960s, particle physicists using atom smashers (particle accelerators and colliders) discovered an array of subatomic particles, with exotic names such as fermions, mesons, hadrons, baryons and leptons. Eventually, there were over two hundred species, a bewildering profusion labelled the 'Subatomic Zoo'.

In the 1970s, the Standard Model was developed, relating the fundamental forces of nature, with the exception of gravity, to the elementary particles. The proliferation of subatomic particles was shown to be the result of combinations of yet more fundamental particles: quarks. Currently, these are regarded as genuinely elementary, in the technical sense of being indivisible – i.e. the smallest units of matter possible. Quarks and associated particles called leptons are collectively termed fermions, and each has a corresponding antiparticle; there are 24 fermions and 24 antifermions. In addition there are 13 bosons, massless particles that mediate forces (including the recently discovered Higgs boson), giving a total of 61 elementary particles.

68 & 70

Atomic weights of missing elements predicted by Mendeleev's periodic law

Russian chemist Dmitri Mendeleev predicted the existence of what were later named gallium and germanium, and even predicted that their atomic weights would be 68 and 70, respectively; their relative atomic masses are now given as 69.723 and 72.63.

When Mendeleev first devised his periodic table of the elements, there were some conspicuous gaps and clashes with accepted data. A fundamental principle of science is that theories must be tailored to fit the experimental evidence; doing it the other way round is seen as a cardinal sin. Yet part of Mendeleev's radical genius lay in his courageous willingness to commit this sin. Where the elements in his table appeared 'out of order', he suggested that their atomic weights had been miscalculated. Where there were gaps, he suggested that the elements that occupied these places simply had not been discovered.

The power of Mendeleev's new periodic law was revealed when he was able to make startlingly accurate predictions about the nature and properties of these missing elements. For instance, there was a gap in the column containing aluminium and indium. Mendeleev predicted the existence of an intervening element with an atomic weight of around 68 and a density of around 6.0 g/cm^3. When this element was identified as gallium in 1875, and its density measured at 4.9 g/cm^3, Mendeleev simply suggested that it be measured again. The correct value turned out to be 5.9 g/cm^3.

74

Proportion of the observable universe made up of hydrogen (%)

By mass, 74 per cent of the observable universe is made up of hydrogen, with about 25 per cent made up of helium. By number of atoms/ions, hydrogen makes up nearly 100 per cent of the universe.

Hydrogen accounts for by far the majority of matter in the observable universe. There are, as a very approximate estimate, around 10^{80} atoms in the observable universe, around 98 per cent of which are hydrogen. The total mass of visible matter is around 10^{52} kilograms, of which about 7.4×10^{51} kg is hydrogen; almost all the rest is helium.

Why the constant use of the qualifier 'observable' or 'visible'? Observations of the movement of matter in and around galaxies suggest that gravitational forces are at work that correspond to an amount of matter ten times greater than the figures given above, but it has so far proved impossible to observe and so is known as dark matter. The true nature of dark matter is unknown but it is possible that some of it could be hydrogen, in which case the amount and proportion of hydrogen in the universe could be still greater.

The abundance of hydrogen, and more specifically the ratio of hydrogen to helium, is a vital clue to the early history of the universe and the most accurate model of the Big Bang (the origin of the universe). The ratio is the product of the speed of expansion of the early universe, and so allows cosmologists to work out exactly how fast it 'inflated' and cooled after the Big Bang.

76

Height of a column of mercury in a barometer as discovered by Torricelli (cm)

Mercury is one of only two elements liquid at room temperature (the other being bromine), and is 13.5 times denser than water, making it an excellent substance with which to make a barometer, as shown by Evangelista Torricelli in 1643.

A barometer is a device for measuring pressure. Most familiar from meteorology, where barometers are used to measure atmospheric pressure (which relates to weather systems and can thus predict imminent weather conditions), this facility is also essential to chemistry, especially pneumatic chemistry or the study of gases. The standard SI unit for pressure is now the pascal, but a plethora of units exists to describe pressure in different fields and aspects. One of the most common and familiar, from its continued use in blood pressure readings and from its historical application as a legacy of the earliest pressure-measuring technology, is the unit based on the height of a column of mercury.

From the inception of the barometer in 1643, mercury was used for registering pressure changes, and even though many pressure meters today measure electronically, pressure readings are still often given as mm/cm Hg, or millimetres/centimetres of mercury. Accordingly, standard air pressure at sea level is usually given as 76 cm Hg. How did mercury come to be associated with the barometer? The answer illuminates how the origins of science owe at least as much to the practical concerns of engineering as to the pursuits of scholars, and reveals how the controversy over the existence of the vacuum was finally settled.

The vacuum problem

A vacuum is the absence of matter. Although most of the universe is a near vacuum, on Earth such a state rarely occurs naturally, although as Torricelli was to demonstrate, it is remarkably easy to create one. The ancient Greeks generally agreed the concept of a vacuum – something occupied by nothing – to be self-contradictory, and dismissed it as an absurdity. By medieval times, the adoption of Aristotelian philosophy as the underpinning of the prevailing societal and religious dogma included the firm belief that, as Thomas Cranmer, the Archbishop of Canterbury, wrote in 1550, 'Naturall reason abhorreth vacuum, that is to say,

Manometers

Barometers are examples of a class of instruments called manometers: devices that use columns of liquid to measure pressure. In 1661, the Dutch scientist Christiaan Huygens adapted Torricelli's liquid column to create the U-tube manometer, where the height of the columns on either side of the U shows the pressure differentials between each side. Combining simplicity with great sensitivity, it went on to become one of the most essential instruments in chemistry, especially for measuring gas pressures.

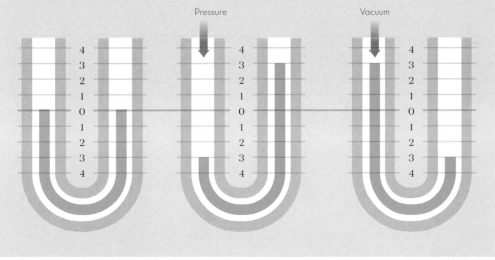

that there should be any emptye place, wherein no substance shoulde be'.

In 1641, Tuscan engineers approached Galileo, now near the end of his life, for help with a practical problem that would prove to have an immediate bearing on the issue of whether or not nature abhors a vacuum. In designing pumps to empty water from mineshafts, the engineers had found that when the column of water in the pump exceeded a height of about 10 metres, it separated from the pump plunger, making it impossible to proceed. Why was this happening, and what was the nature of the space between the top of the water and the bottom of the plunger? Galileo proposed that although nature did abhor a vacuum, this abhorrence could be overcome by the weight of 10 metres of water: the empty space must be the despised vacuum.

The first barometer

Galileo died soon afterwards but shortly before passing away he had taken on a brilliant young secretary, Torricelli, who now set to work to prove his own ideas relating to the pump problem. Working from the widely held knowledge that air had weight, Torricelli correctly surmised that the water in the pump fell because its weight had exceeded that of the column of atmosphere above it. It is this weight, pressing down on the surface of the water beneath the pump, that supports the weight of the water in the pump. To prove his contention, Torricelli built a facsimile of the system in his own house, with a 10-metre tall wooden tube filled with water and topped with a dummy that rose and fell as the level of water in the tube did likewise in response to changes in atmospheric pressure.

A more convincing demonstration, Torricelli realised, could be achieved with something on a smaller, more manageable scale. He knew the volume of mercury to be 13.5 times heavier than an equivalent volume of water, and so he surmised that if mercury were used in place of water in his tube, 13.5 times less would be required to counterbalance air pressure. In June 1644, Torricelli wrote to a friend to explain his methodology: 'We have made

Pascal and the volcano

By 1646 word of Torricelli's experiment had reached the brilliant French mathematician and natural philosopher Blaise Pascal. Pascal was familiar with the elements of the barometer, having earlier demonstrated the principle of hydraulic pressure using a long vertical tube projecting from a barrel. His most brilliant insight was to see that the barometer could be used to measure altitude by monitoring how air pressure declined with height, commissioning his brother-in-law to prove the principle by taking a mercury barometer up an extinct volcano in central France in 1648.

many glass vessels ... with tubes two cubits long. These were filled with mercury, the open end was closed with the finger, and the tubes were then inverted in a vessel where there was mercury ... We saw that an empty space was formed and that nothing happened in the vessel where this space was formed.'

'Many have argued that a vacuum does not exist,' Torricelli wrote, 'others claim it exists only with difficulty in spite of the repugnance of nature [Galileo's position]; I know of no one who claims it easily exists without any resistance from nature.' But it was clear from his experiment that the empty space at the top of the vessel was indeed a vacuum, created by simply turning a tube upside down.

What Torricelli had created was the first barometer. The height of the mercury in the tube fell to a level at which the weight of the liquid remaining in the tube was equal to 'the weight of a column of 50 miles of air' pressing down on the surface of the pool of mercury around the base of the tube. Torricelli found that height to be around 76 centimetres, although the precise level varies with atmospheric pressure.

78.37

Boiling point of ethanol (°C)

Everyone engages daily in what might be termed 'folk chemistry', most commonly in the kitchen. Every time you brew coffee, add salt to boiling water or simply do the washing up, you're harnessing and manipulating chemical processes and phenomena. The fermentation, distillation and application of alcohol is one of the best examples, and utilising the low boiling point of ethanol, the alcohol we drink and cook with, is a case in point.

The boiling point of ethanol is 78.37°C, significantly lower than the boiling point of water and the basis of the technique of 'cooking off' alcohol. The reasoning is that one can add wine, spirits or beer to a dish and easily boil away the ethanol, leaving only the flavours and making the dish safe for drivers, children and non-drinkers.

In fact, cooking or boiling off alcohol is a myth, according to a 1992 study by researchers at the U.S. Department of Agriculture, the University of Idaho and Washington State University. They tested the alcohol remaining in dishes prepared using various cooking techniques and found that, unless the dish is simmered with constant stirring for over half an hour, 45–80 per cent of added alcohol remains after cooking. Simmering with stirring for over 2 hours could reduce that to 5 per cent, but boiling off all the alcohol is almost impossible because alcohol binds with water to form an azeotrope (a mixture of two or more compounds where the ratio cannot be changed by simple distillation). Unless all the liquid is boiled off, some alcohol will always remain.

79

Atomic number of gold

Gold is the 79th element in the periodic table, and probably the one that humans have known about for the longest. Thanks to its unreactive nature, gold can be found as the free metal in nature.

With its striking lustre, high density and resistance to tarnishing or corrosion, gold has long fascinated humans and held great value for them. Archaeological evidence shows sophisticated gold metalwork extant from at least 5000 BC. Initially it was probably obtained from alluvial deposits in riverbeds, but by at least 2000 BC the ancient Egyptians had started mining gold, which can be found in veins of free metal. Gold also occurs in conjunction with silver, quartz and calcite. There is gold in seawater but, at concentrations of around 1 milligram per ton, no one has found an economically viable method of extraction, although Fritz Haber of the Haber process fame did try at one point (see page 10).

Extremely long and incredibly thin

Most of the world's gold is either stored as bullion reserves to back up currency, or made into jewellery. Making very fine jewellery is easy with gold, because it is the most ductile and malleable of all metals. It can be drawn into a thin wire with ease: one gram of gold could be drawn into wire 66 kilometres long, with a thickness of 1 micron, so that it would take just 0.6 kilogram of gold to make a wire long enough to go around the world. It can also be beaten into incredibly thin sheets: a single ounce of gold can be beaten into a sheet roughly 5 metres square. A sheet of gold leaf can be as thin as 0.000127 millimetres, about 400 times thinner than a human hair.

All the gold in the world

Despite the popularity of gold, the ubiquity of gold jewellery and the masses of bullion stored as currency reserves, the total amount of gold ever produced is surprisingly small. According to one estimate, all of the gold that has currently been refined could be placed in a 20-metre cube. According to another estimate, it would fill three and a half Olympic swimming pools. This equates to 170,000 metric tons, the total amount extracted from the Earth as of early 2012. Currently about 2,500 metric tons of gold is mined per year, and two-thirds of all the gold ever taken from the Earth has been extracted since 1950.

This puts into dramatic perspective the astonishing wealth offered by Inca Atahualpa to Pizarro and the conquistadors. He offered to ransom his life by filling the storeroom in which he was being held prisoner to a mark that was as high as he could reach. The dimensions given for the Ransom Room are 22 feet long by 17 feet wide (6.7 metres by 5.2 metres), and the height is recorded as 8 feet (2.4 metres). This works out as 1,630 metric tons of gold, which must have represented a considerable proportion of all the gold ever extracted at that point in history. The conquistadors took Atahualpa's treasure and then murdered him anyway.

▼ Gold bullion; according to a 2011 US Geological Survey report about 37 per cent of extracted gold is in the form of bars or other investment species, with nearly 50 per cent in the form of jewellery.

84

Atomic number above which all elements are radioactive

Elements of atomic number 84 and above are all radioactive because their nuclei are unstable. The 'belt of nuclear stability' ends at element 83 (bismuth).

Among the naturally occurring elements, there are 266 stable nuclides and 65 radioactive nuclides, where a nuclide is a term incorporating all species of atoms or nuclei containing specific numbers of protons and neutrons. Most of the elements have more than one isotope (see page 69), which is why the number of nuclides far exceeds the number of elements. Why are some nuclides stable and others not?

Nuclear stability depends on the ratio between the number of protons (Z) and on the relationship between Z and N, and the overall size of the nucleus (A). Nucleons (nuclear particles, including both neutrons and protons) are bound together by nuclear forces; the more nucleons, the greater the binding forces. But at the same time protons, with their positive charges, repel each other with electrostatic force. So the more protons in a nucleus, the greater the repulsive forces tending to tear it apart and the more neutrons required to boost the binding forces to counterbalance the electrostatic repulsion. The more protons in a nucleus, the more neutrons it needs. Yet having too many neutrons also makes a nucleus unstable, though the reasons for this are not well understood. The balance between these two factors means that the nuclides, when plotted on a graph of Z vs N, fall along a zone known as the belt of stability.

90

Solubility of CO$_2$ in water
at room temperature
(cm^3/ 100 ml)

The solubility of carbon dioxide in water at room temperature and standard pressure is roughly 90 cm^3/100 ml water. The solubility of CO$_2$ in water and how it varies with temperature is of massive importance for the global ecology and has profound implications for the climate.

Carbon dioxide can dissolve in water in straightforward fashion to give aqueous CO$_2$, but this exists in fairly low concentrations. More importantly, the carbon dioxide also reacts with water to produce carbonic acid, a weak acid that dissociates to give excess protons (making the water more acidic and decreasing its pH) and bicarbonate ions:

$$CO_2 + H_2O \rightarrow H_2CO_3 \rightarrow H^+ + HCO_3^-$$

But in seawater the free protons react with carbonate ions (CO$_3^{2-}$) to generate more bicarbonate, giving an overall reaction that sees one carbonate ion consumed for every molecule of CO$_2$ added to the seawater:

$$CO_2 + H_2O + CO_3^{-2} \leftrightarrow 2HCO_3^-$$

This reaction has several consequences. Carbonate ions can react with calcium and magnesium cations in seawater, producing insoluble carbonates that precipitate out. This has led to the formation of extensive deposits of limestone (CaCO$_3$) and dolomite (mixed CaCO$_3$ and MgCO$_3$). Carbonate ions are also critical for marine organisms that build shells, including corals, marine plankton and shellfish. The more CO$_2$ that dissolves in water, the less carbonate is available for shell construction.

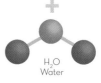
CO$_2$
Carbon dioxide

+

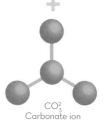
H$_2$O
Water

+

CO$_3^2$
Carbonate ion

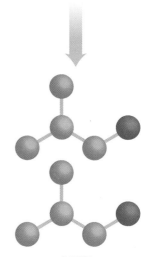
2 HCO$_3^-$
2 bicarbonate ions

▲ The equation showing how carbon dioxide dissolving in seawater reacts with carbonate ions to produce bicarbonate ions, along with simple graphic depictions of the molecules in question.

Ocean acidification

Another way of looking at this is that as CO_2 dissolves in water, it causes it to become more acidic because of the carbonic acid. The massive increase in atmospheric CO_2 from anthropogenic emissions has resulted in increased oceanic absorption and hence lower, more acidic pH in the surface waters of the ocean. Since the beginning of the Industrial Revolution in the second half of the 18th century, the pH of surface ocean waters has fallen by 0.1, representing roughly a 30 per cent increase in acidity. If carbon emissions continue to climb as at present, models show that by 2100 the ocean surface waters could be 150 per cent more acidic, with pH levels not seen in more than 20 million years. This so-called ocean acidification is already having profound consequences on marine life and the marine ecosystem. Organisms that play vital roles at the base of the oceanic food chain will suffer because their shells will effectively be dissolved. For instance, the pteropod, or sea butterfly, is a major zooplankton food source for organisms from krill to whales, including, crucially for human fisheries, young salmon in the North Pacific. Pteropod shells will dissolve almost entirely in the conditions associated with predicted carbon dioxide levels for 2100 under current emission levels. Acidification is already impacting coral reef formation, and may be implicated in the ongoing failure of the oyster harvest along America's West Coast.

▼ Magnified image of a pteropod, showing why it is also known as a sea butterfly.

Warming loop

The solubility of carbon dioxide decreases as temperature rises, and this threatens to cause a feedback loop that might accelerate the greenhouse effect and global warming. As increased carbon dioxide levels lead to global warming, which in turn warms the oceans, so they will be able to absorb less CO_2 and may start to release it, generating positive feedback with increasing temperatures → falling solubility → greater emissions → increasing temperatures.

98

Number of elements that occur naturally on Earth

Although there are currently 118 elements on the periodic table (see page 104), only 98 of them can be found naturally on Earth. This number has been subject to revision.

It has traditionally been said that 91 of the elements can be found in nature: all of the elements up to uranium, atomic number 92, except technetium, number 43. The reason that some elements are not found in nature is that those with the largest nuclei are unstable and radioactive, so that they undergo radioactive decay and transmute into different elements.

Take neptunium, for instance, the next element after uranium. Its most stable isotope, neptunium-237, decays by emitting alpha particles and transmuting into actinium. Its half-life is 2,144,000 years, which sounds a long time but compared to the lifetime of the Earth is relatively short. This means that any neptunium that was incorporated into the Earth when it formed 4.5 billion years ago will have long since radioactively decayed into actinium. This is true of many of the radioactive elements, but clearly not all of them, for uranium, thorium, actinium, protactinium and many other radionuclides are found naturally occurring on Earth. Yet uranium, as we know from its extensive use in fission technologies, is highly unstable and radioactive; how come uranium and these other radionuclides occur naturally on Earth?

The reason is that uranium and the other natural radionuclides have relatively long half-lives. The half-life of uranium-238, the most stable isotope, is 4.5 billion years, the same as the age of the Earth, so roughly half of the U-238 that was present at the formation of the planet is still with us. Even the fissionable isotope, U-235, has a half-life of around 700 million years, so although about 98 per cent of the original amount of this isotope has decayed, there is still enough left for it to be extracted for fission technologies. Elements surviving from the formation of the Earth are known as primordial elements.

Productive pitchblende

It is now believed that in fact there are 98 naturally occurring elements, because many of those radionuclides previously believed only to have been made artificially in atom-smashing laboratories have now been detected in nature. A study in 1971 claimed to have isolated plutonium-244 from Precambrian bastnasite rocks, and although this finding has been regarded with scepticism, it is now known that plutonium, technetium, americium, curium, berkelium and californium can all be detected in very minute amounts in pitchblende, the ore of uranium from which Marie Curie famously isolated radium after much heroic and health-shattering labour (see page 40). They occur because the fission products of natural uranium decay interact with uranium nuclei in the ore to create these short-lived radionuclides, so although they quickly decay they are constantly replenished.

Major players

Although they do thus exist in nature, these elements are present in vanishingly small amounts. In fact, only 83 of the elements occur in significant quantities, and a mere 8 elements are present in the Earth at abundances greater than 1 per cent by mass (in order of abundance: iron, oxygen, silicon, magnesium, nickel, sulphur, calcium and aluminium, with the first four on their own accounting for 93 per cent of the Earth's mass).

104 & 105

Atomic numbers of the elements in the Transfermium Wars

The artificially created elements 104 and 105 in the periodic table are now known as rutherfordium and dubnium, respectively, but their official names were the subject of a fierce US–Soviet battle.

The transuranic elements have all been discovered by smashing together nuclei and nucleons in colliders to create short-lived radionuclides that can be observed just long enough to confirm their existence. The convention for naming such elements is that the discoverers have the right to propose the name, although official adoption of the name depends on the decision of the International Union of Pure and Applied Chemistry (IUPAC).

The names for the transuranic elements up to atomic number 100, fermium, were agreed without dispute, but the first two elements beyond fermium that were isolated, elements 104 and 105, were independently discovered by teams from the Russian Joint Institute for Nuclear Research (JINR) at Dubna and the Lawrence Berkeley National Laboratory (LBNL) in the USA. Each group proposed names with which the other was unhappy, especially when the Russians suggested that 104 should be named after Igor Kurchatov, father of the Soviet atomic bomb. The dispute was known as the Transfermium Wars. Eventually the 'war' was settled by a 1997 IUPAC ruling that accepted the American name for 104, rutherfordium, and gave the name dubnium to 105 in honour of the location of the JINR.

104.5

Angle of H-O-H
bonds in water (°)

The angle between bonds in a molecule is determined by the position of the clouds of electrons emerging from the atoms in the molecule. Electrostatic repulsion drives each cloud to diverge by the maximum distance possible, so that in a molecule such as methane (CH_4), with four electron clouds around the central atom (one for each hydrogen–carbon bond), the clouds arrange themselves in a tetrahedral shape with angles of 109.5 degrees between each bond.

In a water molecule, the oxygen atom also has four clouds of electrons: one for each hydrogen–oxygen bond and two unshared electron pairs. So water should form a tetrahedral arrangement of the four clouds, which would give an H-O-H angle of 109.5 degrees. However, not all four of the electron clouds are equal: because the two electron pairs are very strongly electronegative; they repel each other to give an angle greater than 109.5 degrees, which in turn 'squeezes' the H-O bond clouds together to give an angle of just 104.5 degrees. This gives the molecule 'sides', which in turn makes the molecule polar, with the oxygen 'side' being electronegative and the hydrogen 'side' electropositive. This polarity accounts for the ability of H_2O to form hydrogen or H-bonds: stabilised structures in which a hydrogen atom is in a line between the oxygen atom on its own molecule and the oxygen on another molecule. These H-bonds are the key to the unusual chemistry of water, with profound implications for life on Earth.

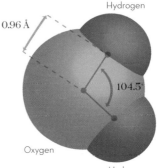

0.96 Å

Hydrogen

104.5°

Oxygen

Hydrogen

▲ 'Soap bubble' schematic of water, showing the distance between each nuclei and the angle formed by drawing straight lines between them.

118

Atomic number of ununoctium, heaviest element ever created

118 is the number of protons in the heaviest element yet created by humans; the atomic mass of the single known isotope is 294, although only a few atoms of the element, which has a half-life of just milliseconds, have been seen.

By firing heavy nuclei at even heavier nuclei at great energies, particle physicists can create new elements. For instance, the Russian scientists at the JINR in Dubna (see page 102), who first produced ununoctium in 2002, fired calcium ions at atoms of californium (itself a synthetic element) for over a thousand hours to produce just a single atom of the new element.

Ununoctium is what is known as a placeholder name. No name has yet been assigned to the element by IUPAC (see page 102), partly because it was synthesised so recently and the evidence for it rests on fewer than a handful of atoms. In the meantime, it is referred to by a name constructed following the conventions laid down by IUPAC. The convention is to use simplifed Latin roots with the standard suffix for an element '-ium'. So ununoctium means literally 'one-one-eighth-ium'.

The initial claim to its creation came from a 1999 paper by researchers at the LBNL (see page 102), but their claims were retracted the following year after failure to replicate. In 2002 it was announced that the claim for discovery of ununoctium had been based on fabricated data. That same year the Russians made some, and in 2006 published their announcement with backing from America's Lawrence Livermore National Laboratory.

126

Atomic number of the 'island of stability'

The 'island of stability' is a realm of the periodic table beyond the heaviest elements yet known, which is believed may exist around atomic number 126.

The stability of an atomic nucleus decreases as the number of protons it contains increases, and although this instability can be counteracted by increasing the number of neutrons, in general nuclei become increasingly unstable as they get bigger. According to one theory, nucleons are arranged in the nucleus in shells, a bit like electrons, and as with electrons, filled shells are more stable. Also, pairs of nucleons are more stable, and these factors combine to produce so-called 'magic numbers' of protons and neutrons that confer enhanced stability on a nucleus. The heaviest stable nucleotide is lead-208, which has magic numbers of both protons and neutrons and so is said to be 'doubly magic'.

Nuclear physicists theorise that superheavy transuranic elements that are doubly magic may buck the trend of increasing instability, and prove to be quite stable, with long half-lives. On a graph of proton numbers versus neutron numbers, these elements would occupy an 'island of stability' amid a sea of instability. A superheavy magic number of protons would be 126, while 184 would be a magic number of neutrons, suggesting that the element unbihexium-310, if it could be synthesised in a suitably powerful particle accelerator, would be doubly magic and possibly have a half-life of days or even years. Such an element could be extremely valuable as a source of neutrons.

▼ Dr Yuri Oganessian of the JINR even suggests there could be a second island of stability around atomic number 164.

140

Temperature threshold for Maillard reactions in cooking (°C)

The temperature above which the Maillard reaction proceeds at a high enough rate to be useful for most cooking is around 140–150°C.

The Maillard reaction may be the most common chemical reaction in human society for it occurs almost every time food is cooked. Named for its discoverer, French biochemist Louis-Camille Maillard, who first characterised it in 1912, the reaction is what happens when sugars and amino acids (of which proteins consist) combine. Although the Maillard reaction occurs even at low temperatures (including inside the human body), for culinary purposes the threshold above which it becomes immediately noticeable and useful is around 140°C.

Flavour saviour

Commonly known as the 'browning reaction', the Maillard reaction is actually a whole series of changes that might more usefully be termed the 'flavour reaction'. It is responsible for many of the colours, flavours and odours associated with a range of foods, from baked bread, biscuits and popcorn to soy sauce, roast coffee and browned meat. In the Maillard reaction, an amino group combines with a sugar to produce water and an unstable intermediate called a glycosylamine, which then rearranges to give a series of aminoketose compounds. These undergo a range of reactions to produce hundreds of different possible

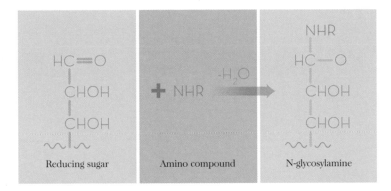

$$HC{=}O \qquad + \; NHR \qquad \xrightarrow{\;-H_2O\;} \qquad \begin{array}{c} NHR \\ | \\ HC{-}O \end{array}$$

Reducing sugar	Amino compound	N-glycosylamine

◀ Typical example of the initial stages in a Maillard reaction, showing a sugar reacting with a generic amino compound (NHR) and losing a molecule of water to give a glycosylamine.

products, including potent scent and flavour molecules and dark pigments. For example, one possible product, 6-acetyl-1,2,3,4-tetrahydropyridine, is responsible for the biscuit or cracker-like odour present in baked goods such as bread, popcorn and tortilla products, while another, 2-acetyl-1-pyrroline, helps give basmati rice its flavour and scent. In both compounds, concentrations of less than 0.06 nanograms per litre (little more than one part in 100 trillion) are needed for a human to smell them.

Wet, wet, wet

In cooking, temperature is the most obvious variable in producing the Maillard reaction. For instance, meat cooked at around 100°C will not go brown and will have little flavour or odour. Moisture content is also important; while water is still present, the temperature of the food cannot exceed the boiling point of the water, and since water is produced in the first step of the Maillard reaction, removing it helps progress the reaction. High temperatures help by boiling off water and drying out the parts of the food to be cooked. However, the Maillard reaction often goes hand in hand with another food reaction, caramelisation, and above around 200°C this takes over, along with simple burning. High pressure raises the boiling point of water so cooking under pressure makes it possible to get Maillard effects even in very wet foods. Alkaline conditions also help, so bicarbonate of soda is often added to help progress Maillard effects.

146

Neutrons in the heaviest isotope of uranium

The atomic mass number (A) of an isotope is the sum of its proton or atomic number (Z) and the number of neutrons (N). Uranium has an atomic number of 92, i.e. it has 92 protons. Different numbers of neutrons in a uranium nuclide give different isotopes. The most important fissile isotope (i.e. the one that can be used for fission) is U-235, which has 143 neutrons (see page 118), but because of its short half-life it is rare in nature (see page 100). Nuclides of uranium-238, the heaviest known isotope of uranium, contain 146 neutrons. U-238 is also radioactive but is several orders of magnitude more stable, with a half-life of 4.5 billion years, meaning that about half the amount of U-238 present at the formation of the Earth is still here, so this isotope accounts for 99.3 per cent of the natural abundance of uranium.

The overwhelming predominance of U-238 in uranium ore posed a massive challenge to the Manhattan Project, the Second World War programme to develop a fission bomb, because scientists needed to extract the minute traces of the lighter isotope. The slight difference in mass between the two isotopes suggested that one way to separate them might be on the basis of this physical property, and centrifugal separation is today the primary method of uranium enrichment (increasing the relative abundance of U-235). U-238 can become fissile by capturing a neutron and transmuting into plutonium-239, a property the next generation of nuclear reactors may seek to exploit.

159

Contents of a barrel of oil (l)

A standard barrel of crude oil, aka petroleum, contains 159 litres. Thanks to some clever chemistry, this 159 litres yields about 170 litres (45 US gallons) of refined products.

The world's most traded commodity is petroleum or crude oil, an umbrella term for liquid fossil fuels that can vary widely in content, properties and appearance. Crude oil is a complex mixture of many different components, mainly hydrocarbons. By weight, about 93 97 per cent will probably be carbon, with most of the rest being hydrogen, along with small amounts of nitrogen, oxygen and sulphur (although the latter can account for up to 6 per cent in some crudes), and traces of metals. The hydrocarbons fall into four classes: paraffins (aka alkanes – the most valuable class; 15-60 per cent), naphthenes (30-60 per cent), aromatics (3-30 per cent), and small quantities of asphaltics. The precise composition depends on the location of the oil reservoir.

From crude to refined

Crude oil can vary in colour from black to straw-coloured, and in consistency from very light and highly volatile to treacly, tarry substances that are almost solid. Light crude is more valuable because it is easier to extract useful and valuable petroleum products. Oil refining is the set of processes by which crude oil is separated into different components or fractions and these in turn are processed into usable end products.

■ Diesel 12
■ Heating oil 1
■ Jet fuel 4
■ Other products 6
■ LPG 2
■ Heavy fuel oil 1
■ Gasoline 19

▲ Graphic showing the products of distillation and cracking as a proportion of the original barrel of crude oil, with quantities in gallons, according to the US Energy Information Administration.

The first step in oil refining is to heat the crude oil and use the different weights and boiling points of the various fractions as the key to their separation: heated crude oil is passed into a distillation unit, a tower in which the lightest fractions, including gasoline and liquid petroleum gas (LPG), rise to the top and the heavy ones, known as gas oils, sink to the bottom, while medium-weight fractions, including kerosene and diesel, are in the middle. At each level, the fraction is drawn off. Gas oils are then subjected to high heat and pressure in the presence of catalysts in order to crack them, which involves breaking down long-chain hydrocarbons into smaller ones, and these are then treated further with alkylation and other processes that build them up again into useful end products such as gasoline. This secondary process is known as conversion. The end products of oil refining include LPG, gasoline, jet fuel, heating oil and diesel. Because cracking turns dense, heavy hydrocarbons into lighter ones, the volume of the products actually exceeds the volume of crude oil put in. The yield of 170 litres of petroleum products from a barrel of crude oil includes around 72 litres of motor gasoline and 38 litres of diesel.

Proliferate petrol

Petroleum products extend far beyond different types of fuel. Crude oil distillates are the basis for much of the plastic industry and also find their way into an extraordinary range of products. A non-exhaustive list includes: tyres, crayons, heart valves, ammonia, washing-up liquid, crayons, spectacles, deodorant, ink, computers, CDs and DVDs.

169

Weight of Jan van Helmont's willow tree (lb)

169 lbs was the weight of a willow tree as measured by the 17th-century alchemist Jan Baptista van Helmont, after he had grown it from a sapling and fed it only water for five years, in the first rigorous experimental exploration of photosynthesis.

Van Helmont was a Flemish nobleman who devoted his life to the pursuit of natural philosophy, including alchemy, for which he would run into trouble with the authorities. His researches were not published until 1648, four years after his death. The most celebrated of his experiments was an extension of one carried out by Nicholas of Cusa in the 15th century. Nicholas had noted the tiny increments in weight of a plant growing in a pot (a closed system), the first research to suggest that plants could somehow gain matter from nothing more than air and water.

Out of water only

Van Helmont did a more rigorous version of this experiment. 'For I took an Earthen Vessel, in which I put 200 pounds of Earth that had been dried in a Furnace ... and I implanted therein the ... Stem of a Willow Tree, weighing five pounds and about three ounces...' Covering the vessel with a shield of tin, he fed it nothing but distilled water.

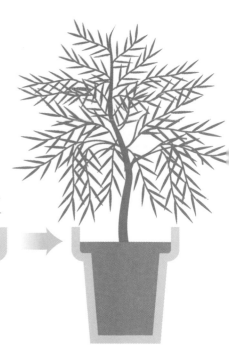

▼ Van Helmont's willow tree at the start of his experiment and after five years of receiving only water.

Five years later, he dug up what was now a small tree and weighed it: it now weighed 169 lbs. He also dried and weighed the earth in the pot, finding that it still weighed almost exactly 200 lbs. 'Therefore,' he surmised, '164 pounds of Wood, Barks, and Roots, arose out of water only.'

It seemed that somehow the willow tree had transformed water into plant matter. This made perfect sense to van Helmont, who had formed a strong belief in water as the fundamental element, following the theory of the ancient Greek philosopher Thales (see page 54). 'The whole rank of Minerals,' he opined, 'do find their Seeds in the Matrix or Womb of the Waters.'

On the trail of CO_2

In fact, the primary source of the willow tree's weight gain was not the water but the air, specifically carbon dioxide in the atmosphere. The process of photosynthesis ('making with light') combines CO_2 with H_2O to produce carbohydrates and oxygen, in a process powered by sunlight (see page 133). Ironically, van Helmont, despite failing to identify carbon dioxide as the real source of the willow's increase in mass, was the first man to recognise the likely existence of the gas. Experiments in which he dissolved metal in acid and then recovered exactly the same weight of metal had convinced him of the principle of conservation of mass; i.e. that matter could not be created or destroyed, only changed in nature. Noting that after burning 62 lbs of charcoal he was left with 1 lb of ash, van Helmont theorised that the other 61 lbs of matter had escaped as some sort of vapour or airy spirit: 'I call this Spirit, unknown hitherto, by the new name of Gas, which can neither be constrained by Vessels, nor reduced into a visible body.' The gas given off from burning charcoal he called *spiritus sylvester* ('spirit of the wood'); it would be identified as a new air by Joseph Black (see page 36), and eventually become known as carbon dioxide.

169.8

Molecular weight of silver nitrate (g/mol)

169.8 grams is the weight of a mole of silver nitrate ($AgNO_3$), a relatively stable compound of silver, from which the metal can readily be mobilised for uses including photography.

Silver forms a number of compounds with the marvellous property of readily being reduced to metallic silver, causing a previously whitish or transparent substance to darken noticeably. This peculiar phenomenon had been known for some time when, in the 1720s, German physicist and physician Johann Heinrich Schulze became the first man to use it for photography – 'writing with light'. Schulze had experimented on the compound silver nitrate, baking it in an oven to prove that while this did not lead to discoloration, exposure to sunlight did. Using stencils laid on top of jars of chalk mixed with silver nitrate. 'The sun's rays,' Schulze noted, 'where they hit the glass through the cut-out parts of the paper, wrote each word or sentence on the chalk precipitate so exactly and distinctly that many who were curious about the experiment but ignorant of its nature took occasion to attribute the thing to some sort of trick.'

▲ A vial of silver nitrate, which forms white sugar-like crystals. In solution it can cause skin staining and, with prolonged exposure, burns.

The colour of nitrate

Silver nitrate would also be used in the first, halting attempts to create a photograph in the modern sense of the word: a picture made with light. Inspired by the work of Schulze, and follow-up experiments on silver salts by Carl Scheele, the English physicist

Thomas Wedgwood and his friend Humphry Davy investigated photography in the late 18th and early 19th century. Silver nitrate proved too insensitive to record images projected by a camera obscura (a darkened room or box into which an image is projected by a lens gathering light from outside), but when they tried contact printing by placing a painting on glass on top of a sheet treated with silver nitrate, they found that exposure to sunlight resulted in 'distinct tints of brown or black, sensibly differing in intensity according to the shades of the picture, and where the light is unaltered, the colour of the nitrate becomes deepest'. But the simple act of looking at their photograph meant exposing it all to the light, and since they could not 'fix' the unexposed silver salt (i.e. prevent it from reducing to metallic silver), the whole sheet quickly darkened and the image faded.

Silver nitrate seemed not to be the best compound to use for true photography. Silver halide salts, such as silver chloride or silver iodide, are more sensitive and thus more suitable, but this sensitivity is also one of the features that makes them hard to use in practice. To prepare a plate for photography, the silver salt must be applied as a thin film, but silver halides are insoluble, so it is difficult to do this; and the reactivity that makes them so

Ag   Cu   HNO₃

Real-life subject   Camera

Basic plate   Plate is trimmed   Polishing   Sensitization
               and edges bent

light-sensitive also means they cannot be stored for long periods. The solution to these problems turned out to be silver nitrate, which thus still had a vital role to play in early photography.

Wet plates

The technology that eventually developed from the daguerreotype invented by Louis-Jacques-Mandé Daguerre and Joseph Nicéphore Niépce was the 'wet plate' process. A cameraman took with him a small mobile laboratory, allowing him to prepare plates just before use. A glass plate was washed in a solution of a stable salt such as potassium bromide. The photographer then made a 'bath' of silver nitrate, which is stable enough to be stored for long periods and lugged around, is easily soluble, and readily displaces other metals from their salts. When the plate was dipped into the bath, the silver in the nitrate displaced the potassium in the bromide, forming insoluble silver bromide, which then precipitated out onto the glass plate as a thin film. The plate was now ready for immediate use, and photography could commence.

The apparatus necessary to produce such plates included heavy chemicals, a portable dark room in which the solutions and

▼ The somewhat laborious daguerreotype process, showing a silver-plated copper plate being trimmed and bent at the edges, polished with corrosive nitric acid, sensitised with iodine fumes to create a reactive silver halide coating, exposed to light reflected from the subject (in this case Daguerre himself) and then developed with mercury fumes, fixed with sodium thiosulphate to remove any remaining halide, and, in a later development, gilded with gold chloride over heat to protect the fragile silver dust image.

$Na_2S_2O_3$ ·············· $AuCl$

Plate is exposed   Development   Fixing   Gilding

Hg

plates could be prepared and then developed, and the camera itself – at this time a large unwieldy apparatus, which had to be mounted on a tripod. The chemicals involved, especially the silver salts, were messy and toxic. Silver salts mark the skin indelibly, with brownish-black stains, and silver nitrate is highly toxic; the stains would soon turn into burns if exposure was not controlled. Silver nitrate also irritates the eyes and can cause blindness. If ingested or inhaled, silver nitrate is not expelled from the body but lingers, slowly being converted to silver sulphide, which is black, leading to the appearance of dark stains all over the body, a condition known as argyria.

The wet plate process of photography was superseded in 1880 by George Eastman's Kodak process, which used gelatine to produce 'dry' plates. Eight years later, Eastman introduced flexible film, and photography became dramatically more accessible. Silver nitrate remains to this day an integral part of the process of producing photographic film. Whenever silver salts are involved, silver nitrate is preferred as the first reagent, from which halide salts are prepared nearer the end of the process.

▲ A vintage wet-plate camera. The length of exposure would be controlled, not with a shutter as in a modern camera, but simply by removing and replacing a cap over the lens piece.

Mirror, mirror

Another application of silver nitrate is in the 'silvering' process used to make mirrors. Silver is the best-known reflector of visible light, but because it oxidises quickly and darkens, its reflective surface must be sealed off from air. One side of a piece of glass is washed with silver nitrate solution, and then sodium hydroxide and ammonia are added to produce a silver–ammonia complex. Adding sugar reduces the silver to its metallic form, which is now deposited in an even layer on one side of the glass, turning it into a mirror.

235

Atomic mass of the isotope of uranium used in making the first atom bomb

The fissile naturally occurring isotope of uranium used to create the first atomic bomb was U-235, which has 92 protons and 143 neutrons, giving it a relative atomic mass of 235.

The story of the atomic bomb begins with the discovery of radioactivity, which gave the first clues that the atom is divisible and that tremendous energies are involved in the related processes. As early as 1914, H.G. Wells's prophetic novel *The World Set Free* described atomic bombs of devastating power. Turning science fiction into fact would have to wait more than 30 years, a long journey involving many milestones.

The discovery of fission

Among the first of these was James Chadwick's 1932 confirmation of the existence of the neutron, the neutral nucleon long suspected by Ernest Rutherford (see page 26). Discovery of the neutron prompted Hungarian physicist Leó Szilárd to conceive of a nuclear chain reaction. Szilárd recognised that the neutron, with no electrostatic charge, can more easily approach other nuclei and fuse with them, triggering nuclear instability that may result in the release of further neutrons, setting in motion a chain reaction that would liberate tremendous amounts of energy. Such a chain could only be set off by a nuclear reaction that resulted in 'free' neutrons, but where could such a reaction be found?

The answer emerged with the discovery of nuclear fission, the phenomenon whereby an unstable nucleus splits in two, rather like an amoeba dividing to produce two daughter cells. The discovery of the neutron had prompted many attempts to create transuranic elements and new radioisotopes by bombarding uranium with neutrons. In the 1930s it was assumed that a nucleus could only change size in small increments, by absorbing neutrons or ejecting alpha particles (consisting of 2 neutrons and 2 protons). Hence the German-Austrian nuclear chemists Otto Hahn and Fritz Strassman and their collaborator Lise Meitner struggled to explain results from their experiments on uranium, which seemed to show that its products included isotopes of barium, element 56, with masses of just 140 or so.

In late 1938, Meitner and her nephew Otto Frisch realised that the production of barium could be explained by the splitting of the uranium nucleus, a process they called fission. News of this breakthrough reached America a few weeks later, after Frisch had personally relayed it to the Danish scientist Niels Bohr just before he boarded a ship heading for a conference in Washington. Fission became the subject of feverish research activity, and it was soon shown that uranium fission could produce free neutrons. A chain reaction might be possible, and Szilárd and others worked to mobilise scientific and political impetus for research into producing chain reaction technology.

▲ Otto Hahn and Lise Meitner at work in their laboratory, before Nazi persecution drove Meitner into exile.

The fissile isotope

Among the many obstacles such a goal would have to overcome was the problem of the apparently temperamental nature of uranium. One reason fission had not been spotted earlier is that uranium rarely demonstrates the phenomenon. On the whole, uranium nuclei seemed to respond to neutron capture by beta decay, in which the neutron emitted a beta particle (an electron) and became a proton, so that the uranium transmuted into element 93 (later named neptunium). It was Bohr who showed that the rarity of fission is down to the fact, only discovered in 1935, that uranium is a mixture of two isotopes, U-238 and U-235.

The lighter isotope is much more unstable and thus much more likely to respond to neutron capture by undergoing fission, but this instability means that it is very rare in nature because most of it has decayed to different elements. The natural abundance of U-235 is just 0.7 per cent.

A fission bomb made with natural uranium is possible but huge quantities would be needed to achieve critical mass and the explosive yield would be low. In February 1940, Frisch worked out that if a bomb were made of pure U-235, the critical mass would be kilograms rather than tonnes. It now became apparent that the crucial challenge facing the effort to create an atomic bomb would be the enrichment of uranium to increase the proportion of U-235. Since the two isotopes of uranium are chemically identical, there are no simple chemical methods of isolation/enrichment. Instead, enrichment depends on the physical differences between the isotopes, namely their relative atomic masses. The Manhattan Project, the colossal engineering endeavour to create the atomic bomb, inaugurated in 1942, pursued multiple enrichment strategies in tandem.

▼ Nuclear fission: addition of a neutron to a U-235 nucleus causes it to split into two 'daughter' nuclei, at the same time releasing energy and two more neutrons.

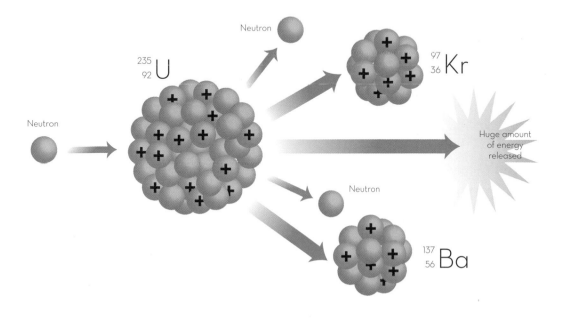

Neutron

$^{235}_{92}$U

Neutron

$^{97}_{36}$Kr

Huge amount of energy released

Neutron

$^{137}_{56}$Ba

The enrichment challenge

The first successful isolation of U-235 was achieved in April 1939 by Alfred Nier, using the same technology that had been used to discover the isotope: a mass spectograph, which uses an electromagnetic field to deflect beams of different isotopes by differing degrees. Electromagnetic principles would also form the basis of the first large-scale attempt to engineer an enrichment technology: Ernest Lawrence's work at Berkeley in California, in which he converted a cyclotron particle accelerator into an isotope separation device he named the calutron. By February 1942, the calutron method appeared to be the most promising, but in the long term it proved prohibitively inefficient and expensive, costing millions of dollars to produce just a gram of U-235.

At Oak Ridge in Tennessee, the Manhattan Project built vast engineering plants to pursue other techniques of isolation and enrichment. The most successful was the gaseous diffusion technique, which works on the basis that less massive gas particles find it easier to diffuse through a porous membrane. Uranium atoms were mobilised into gaseous form by creating uranium hexafluoride, and this was allowed to diffuse through porous membranes to enrich the proportion of the lighter U-235 hexafluoride. The highly corrosive nature of hexafluoride and the need to avoid contamination led to the development of new materials such as Teflon, and meant that the K-25 plant constructed to pursue gaseous diffusion enrichment on a massive scale eventually cost hundreds of millions of dollars, making it the most expensive part of the Manhattan Project.

The Manhattan Project eventually relied on a combination of enrichment techniques including both gaseous diffusion and calutrons, and by the beginning of August 1945, enough U-235 had been accumulated to create Little Boy, the first atomic bomb, which would be dropped on Hiroshima on 6 August.

273.15

Freezing point of water on the
Kelvin temperature scale (K)

There are three different temperature scales in common use throughout the world. The first standardised temperature scale was created by 18th-century German physicist Daniel Gabriel Fahrenheit. He set the zero point of his scale as the coldest temperature he could achieve, a mixture of ice and salt, and then gave arbitrary values of 30 and 90 to the freezing point of water and human body temperature, later revised to 32 and 96 (although subsequent measurements refined the latter reference point to 98.6). Eventually, the Fahrenheit scale had the freezing and boiling points of water at 32 and 212 degrees respectively, giving a span of 180 degrees familiar to scientists. In 1742, Swedish astronomer Anders Celsius designed a metric scale of 100 degrees, often described as centigrade.

In the early 1800s, the relationship between the temperature and volume of a gas was quantified, allowing scientists to work out that an ideal gas will have zero volume at a temperature of 273.15 below zero centigrade. William Thomson, Lord Kelvin, used this to establish an absolute temperature scale in place of the relative ones of Fahrenheit and Celsius. Absolute zero is the theoretical point at which a system has no energy and particles are motionless. On this scale, the freezing/melting point of water is at 273.15. The units in the Kelvin scale are now called kelvins, and because it is an absolute scale they do not have degrees; however, they are equal in magnitude to degrees Celsius.

274

Number of plates in the voltaic battery used by Humphry Davy to isolate potassium

In 1807, Humphry Davy used one of the largest voltaic batteries yet created to electrolyse the molten salts of potassium and sodium, thus becoming the first to isolate these alkaline earth metals and discover two new elements.

Antoine Lavoisier had speculated that the alkaline metal oxides potash and soda would prove to contain two new elements, but the technology to decompose these alkalis into their component elements did not yet exist. In 1799, this technology was invented by the Italian Alessandro Volta, who, building on the work of his countryman Luigi Galvani, created the first electric cell or battery in the form of his eponymous pile. A voltaic pile interleaved alternating discs of zinc, silver, copper or other combinations of conductive metals with cardboard pads soaked with brine. Electrons moving through the conductive brine electrolyte flowed from one metal disc to the next, generating an electrical current when the poles of the cell were connected to complete a circuit. The currents produced were relatively weak, but the more pairs of metal included in the pile, the greater the force of electricity that could be produced, and by 1805 Volta was calling for batteries of 1,800–2,000 pairs.

Volta's pile design was vertical, with a stack of discs held in a cage of vertical rods. Not long after news of the pile was disseminated, the Scottish chemist William Cruickshank came up with a horizontal design, where metal plates were slotted into a long wooden trough insulated with resin. Here they could sit in a bath of electrolyte. Much easier to extend to massive numbers of metal plate pairs, the Cruickshank design became the standard form of electrochemical battery for the next 35 years.

'The highest electrical power I could command'

In 1800, Humphry Davy, then a young tyro at the Pneumatic Institute in Bristol (see page 83), quickly acquainted himself with the possibilities of the new technology. At the Royal Institution in London seven years later, he made use of the establishment's set of powerful batteries to achieve the electrolytic decomposition Lavoisier had foreseen. In his written version of a famous lecture he gave in 1807, Davy explained, 'I acted upon aqueous solutions of potash and soda, saturated at common temperatures, by the highest electrical power I could command and which was produced by a combination of Voltaic batteries belonging to the Royal Institution, containing 24 plates of copper and zinc of 12 inches square, 100 plates of 6 inches, and 150 of 4 inches square, charged with solutions of alumn and nitrous acid.'

Joy uncontained

Davy connected the electrodes of this combined battery of 274 plates to a spoonful of molten potash (potassium oxide), and observed as 'aerifom globules, which inflamed in the atmosphere, rose through the potash'. Davy 'could not contain his joy — he actually bounded about the room in ecstatic delight' The experiments to isolate potassium, and subsequently sodium from soda, exhausted the RI batteries, and a public subscription to upgrade them led to Davy being provided with a battery of 2,000 pairs of plate; its total active electrode area was a colossal 80 m^2. Powerful forces had been unleashed.

334

Latent heat of fusion
of water (kJ/kg)

A kilogram of liquid water at 0°C fusing or freezing into a kilogram of ice will release 334 kilojoules of energy; this is its latent heat of fusion.

Latent heat is heat energy that does not produce a change in temperature. It is a slippery concept to grasp because of the intuitive assumption that putting energy into a system must result in an increase in temperature, and vice versa. In fact, changes of phase absorb or release heat without affecting the temperature. This counter-intuitive phenomenon was first explained by the Scottish physician and chemist Joseph Black, who observed that, contrary to the unexamined assumptions of the time, ice heated to 0°C does not melt all at once with the addition of a bit more heat, but melts slowly, while as it does so its temperature remains at 0°C.

Experimenting with blocks of ice dropped into hot water, Black showed that ice can absorb a surprising quantity of heat without any increase in temperature, and that even more heat 'disappears' when liquid water changes phase into steam. The heat, he wrote, 'appears to be absorbed and concealed within the water, so as not to be discoverable by the application of a thermometer ... it is concealed, or latent, and I give it the name of latent heat'.

Careful measurements revealed that the latent heat of fusion (the heat released when water freezes into ice, or is absorbed when ice melts into water) is 334 kJ/kg, while the latent heat of vaporisation or condensation of water is 2,260 kJ/kg.

356.7

Boiling point of mercury (°C)

Mercury, one of only two elements that are liquid at room temperature, boils at 356.7°C, meaning it is liquid over a relatively wide temperature range.

With a freezing point of -38.9°C, mercury was known to the ancients in its pure liquid form. Despite its metallic lustre, which led to its common names of hydrargyrum ('water silver') and quicksilver, it was not considered to be a metal; but after the arctic Russian winter of 1759-60 made it possible for Professor Braune of St Petersburg to freeze the element, its metallic character was widely acknowledged.

The wide temperature range over which mercury stays liquid makes it an excellent medium for thermometers and barometers, which retain their measuring ability even in extreme conditions. The high boiling point of mercury also engendered a brief vogue in America from the 1920s–40s for mercury-vapour driven turbines for generating electricity. Using a medium with a high boiling point offers much greater levels of efficiency than steam-driven turbines. According to a report in *Popular Science* in March 1931, a mercury vapour plant in South Meadow, New Hampshire, produced 143 kW-h of electricity per 100 pounds of coal burned, while the best conventional steam plant produced only 112 kW-h and the average for coal was just 59 kW-h. However, mercury vapour and mercury compounds are highly toxic, making mercury-vapour power plants a health and safety nightmare.

535

Wavelength of characteristic green line in the emission spectrum of thallium (nm)

When thallium is heated in a flame, like all elements it gives off light with a distinctive spectrum. Viewed with a spectroscope, this spectrum has a characteristic green line representing light with a wavelength of 535 nanometres.

The spectroscope was an exciting invention for analytical chemists after the great German team of physicist Gustav Kirchhoff and chemist Robert Bunsen developed it (see page 129) and showed how it could be used to read the emission spectra of substances heated in a flame. Emission spectra reveal what elements are present even when chemical analysis cannot, and it was soon seized upon to prove the existence of several new elements: caesium, rubidium, indium and thallium.

Green shoots

This last element, a metal similar to lead, was discovered by the English scientist William Crookes in 1861. Investigating the emission spectra of deposits recovered from the chimney of a sulphur factory, Crookes noticed a distinctive green line. It reminded him of the green of new shoots in spring, so he named the new element thallium (from the Greek for a green shoot). Crookes struggled to isolate more than a few grams of the new element, while Frenchman Claude-Auguste Lamy, who independently discovered the element in 1862, managed to obtain an ingot of the metal. A tussle over priority ensued, but Crookes

The poisoners' poison

Thallium has enjoyed a dramatic career as a poison. It is highly toxic to animal physiology because cells easily mistake it for potassium ions, leading to massive disruption of metabolic processes, especially in nerve cells, heart muscle and hair follicles. It was widely used as a rat poison, and its hair-loss effects led to the popularity of thallium-rich depilatory creams, which contained a fatal dose in each tube. More sinisterly, thallium became known as 'the poisoners' poison', thanks to its popularity with murderers and assassins. Among the most notorious were Saddam Hussein's half-brother Barzan Tikriti, feared head of Hussein's secret police, who used thallium to kill dissenters, and Graham Young, the Teacup Poisoner of 1960s and 70s Britain.

prevailed, helped by his success in determining the atomic weight of the new element, in the process of which he started down the road of his distinguished career.

The light windmill

In order to accurately weigh minute quantities of thallium, Crookes employed a vacuum balance: a set of scales inside a partially exhausted bulb, in which the near vacuum helped to avoid gas molecules skewing the measurements. Noticing that incident light appeared to affect the scales, Crookes was inspired to create the radiometer, or 'light windmill', lauded by electrical inventor Nikola Tesla as 'the most beautiful invention' ever. A radiometer consists of four vanes, each blackened on one side and reflective on the other, which spin around a vertical axis inside a glass bulb from which most of the air has been expelled (known as a partial vacuum). Shining a light on it causes the vanes to rotate, as if the light were somehow pushing them around. In fact, the rotation is linked to local heating of the rarefied gas around the vanes, and a more complex phenomenon called thermal transpiration.

589

Wavelength of Fraunhofer's D line in the spectrum of the Sun (nm)

That light could be split into a spectrum of coloured bands by passing it through a prism has been known since at least the Middle Ages. Isaac Newton used the phenomenon to show how white light is a mixture of different colours in a famous demonstration he termed the Experimentum Crucis, in which he used one prism to split white light into its component colours and then another to recombine them into white light. In 1802, English scientist William Hyde Wollaston observed more closely the spectrum of sunlight passed through a prism and noticed it to be crossed by several dark lines. This marked the birth of solar spectroscopy, or perhaps more accurately, the start of its gestation.

Fraunhofer's lines

During the Napoleonic Wars, German lens-maker Joseph von Fraunhofer was looking for a way to calibrate the prisms and lenses he produced, for standardisation purposes. In 1814, revisiting Wollaston's discovery, Fraunhofer started to produce a detailed map of the lines in the solar spectrum, labelling each with a letter, assigning 'D' to a particularly notable dark line at 589 nm. Fraunhofer's lines proved to be an invaluable tool for calibrating prisms, but in terms of an explanation remained a curio for a further 45 years.

▲ Portraits of Gustav Kirchoff (above) and Robert Bunsen (below), German scientists whose fruitful collaboration provided an early model of the power of an interdisciplinary research team.

Emission matches absorption

In 1849, Leon Foucault noticed that the emission spectrum of a sodium lamp produced strong bright lines, including one that matched, in terms of its position in the spectrum, the dark D line in the solar spectrum. In the 1850s, Robert Bunsen and Gustav Kirchhoff at the University of Heidelberg created a sophisticated spectrometer by combining in one box a prism, lens and collimator, allowing them to make accurate measurements of both emission and absorption spectra. The former are produced when a heated gas emits light, and the latter when light is passed through a heated gas and some is absorbed. Investigating Foucault's finding, Kirchhoff and Bunsen found that the sodium emission line did indeed match exactly with the solar D line, and they were further able to show that a hot gas absorbed light at the same wavelengths at which it emitted light. This explained the D line: if hot sodium gas emits light at the D wavelength, the solar D line must be evidence that hot sodium gas between the Sun and the Earth is absorbing that wavelength. In other words, the D line is evidence that the atmosphere of the Sun contains sodium. In a paper of 1859, the two Germans wrote, 'the dark lines of the solar spectrum which are not caused by the zterrestrial atmosphere, arise from the presence in the glowing solar atmosphere of those substances which in a flame produce bright lines in the same position.'

Bunsen and Kirchhoff went on to find evidence for the presence of iron, magnesium, sodium, nickel and chromium in the atmosphere of the Sun. Their spectroscope had kicked off a new age in astronomy, enabling analysis of the chemical makeup of bodies hundreds of millions of miles away. Solar spectroscopy would soon lead to the discovery of the element helium, while on Earth spectroscopy identified a number of new elements (see thallium, page 126).

▼ Light from the Sun, refracted through a prism, reveals a spectrum with characteristic dark bands corresponding to light-absorbing elements in the solar atmosphere.

600.61

Melting point of lead (K)

Lead is sometimes found in nature as the free metal but is mostly recovered from ores, as was the case in ancient times. It was during the Roman era that lead production increased significantly. With its low melting point, lead could be melted on a camp fire, and when solid it was easy to work into sheets and pipes, while its resistance to corrosion made it extremely useful. It could be alloyed with tin to make pewter, and as white lead, or ceruse, it was a popular pigment for decoration and cosmetics. The Romans developed lead mines across the breadth of their empire, from Greece and Sardinia to Spain and Britain, producing over 100,000 tonnes of lead a year. Ice cores and samples from peat bogs allow archaeologists to reconstruct ancient levels of atmospheric contamination, showing that around this time the level of lead pollution increased from 0.5 parts per trillion (ppt) in prehistoric times to 2 ppt in the Roman era.

Sweet poison

Lead is extremely toxic, with a concentration of just 80 micrograms per 100 millilitres in the bloodstream causing acute lead poisoning. Chronic poisoning is linked to much lower levels still, and extensive use of the metal has been controversially linked to the decline of the Roman Empire. This mass of lead posed a potential health problem of critical dimensions.

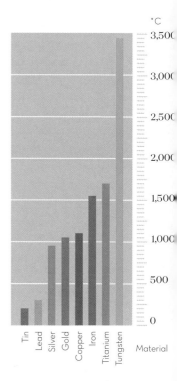

▲ The melting points of various metals compared.

The Romans used lead to make corrosion-resistant pipes for their hydrological engineering, but this could potentially have caused lead to leach into their water supply. Mildly acidic rainwater could well have dissolved lead from cisterns and pipes at toxic levels. The Romans may even have deliberately added lead to their food in the form of *sapa*, a sweetener made by boiling down wine remnants into a syrup. The best recipes called for the cooking of sapa to be done in lead pans, which might have resulted in the production of high levels of lead acetate, aka lead sugar, adding to the sweetening effect. Modern reconstructions suggest that sapa might have contained up to 1,000 parts per million of lead (0.1 per cent), meaning that a single spoonful could result in mild poisoning.

Decline and fall

Figures like these suggest that the ancient Romans might have suffered from widespread chronic lead poisoning, which is known to cause fertility and cognitive problems and even mental illness. In 1965, S.C. Gilfillan claimed that 'lead poisoning is to be reckoned the major influence in the ruin of the Roman culture, progressiveness, and genius' and although his argument was demolished at the time, the theory made a comeback in 1983 with a book by geochemist Jerome Nriagu, *Lead and Lead Poisoning in Antiquity*. Most experts discount the theory, however, pointing out that the Romans were well aware of the existence and symptoms of lead poisoning, and that lead water pipes could be dangerous, and the fact that they probably were not exposed to dietary levels as high as Nriagu claimed. For instance, the hard water of Rome deposited a layer of limescale on the inner surfaces of its lead pipes, preventing leaching of lead. Meanwhile, sapa was probably mainly produced in cheaper bronze pans, not lead ones. Low birth rates among the Roman elite long pre-dated the decline of their empire, and there are socio-economic, geopolitical and ecological reasons for its eventual collapse far more compelling than spurious medical ones.

850

Year of discovery of gunpowder

In around AD 850, Chinese alchemists seeking to create the Elixir of Immortality created the first gunpowder mix, an accidental breakthrough in chemistry that would transform world history.

Chinese alchemy dates back to at least the time of the legendary 3rd-century BC first emperor, Qin Shi Huang, who was obsessed with immortality and sponsored alchemists from across his empire to develop the fabled Elixir of Immortality. By the 9th century AD, Chinese alchemists were familiar with sulphur and saltpetre (potassium nitrate, obtained in ancient times from bird droppings or urine). A Chinese alchemical text from around 850 records what happened when alchemists mixed charcoal from burnt willow with honey, sulphur and saltpetre: 'Smoke and flames result, so that their hands and faces have been burnt, and even the whole house burnt down.' The saltpetre had acted as a powerful oxidiser, enabling the carbon to oxidise rapidly and explosively into carbon dioxide, with the sulphur helping to lower the fusion temperature of the mixture and to mobilise the saltpetre, accelerating the speed of burning of the whole mix.

At first, the military applications of the new concoction were not recognised and it was touted as a pesticide or skin treatment, but by the 11th century the Song dynasty was using a range of gunpowder recipes to hold off Mongol invaders, and by the 13th century black powder had reached Europe. Here it would help dismantle the old feudal order by rendering obsolete castle walls and knightly armour.

▲ Black powder, aka gunpowder. The size of the grains helps determine the speed of burning and thus the nature of the resulting explosion.

893.49

Molecular weight of
chlorophyll a (g/mol)

The molecular weight of chlorophyll a is 893.49 grams per mole. It is a large organic molecule comprising a porphyrin with an internal ring of nitrogen atoms that bond to a central magnesium atom. Capable of capturing photons from solar radiation to excite electrons, chlorophyll a is the primary driver of photosynthesis.

Building on Priestley's observation that plants produce oxygen (see page 46), in 1779 Jan Ingenhousz showed that this happened when the green parts are illuminated with sunlight. Later researchers showed that carbon dioxide and water are the other reactants involved in this process, which was eventually, in 1893, labelled photosynthesis. In 1818, the green pigment found in plants was given the name chlorophyll by Pierre-Joseph Pelletier and Joseph-Bienaimé Caventou; 20 years later it was found that this pigment was the agent of photosynthesis.

In 1913 Richard Willstätter and Arthur Stoll showed that chlorophyll is composed of two molecules, labelled chlorophyll a and b. Chlorophyll a is the most abundant. It absorbs light strongly at the blue and red ends of the spectrum, leaving green light to be reflected back, thus accounting for the green colour of plants. Photons of light from the absorbed parts of the spectrum are captured and excite electrons that can then be passed along a chain to carbon dioxide, liberating the oxygen so the carbon can be fixed into energy-rich carbohydrates, which can then be used for metabolic respiration.

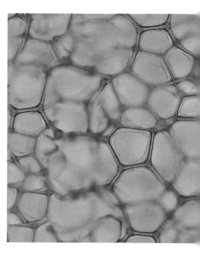

▼ Plant cells, tinged green because the plant's chlorophyll has absorbed much of the red and blue light.

1661

Year of publication of
Boyle's *Sceptical Chymist*

Robert Boyle was an Anglo-Irish aristocrat from a wealthy family, who pursued researches in natural philosophy. He made important discoveries such as a colour test for acids, and worked on air pumps, formulating the first of the gas laws (see page 62), now known as Boyle's Law: 'compressing a volume of gas by half will cause the pressure to double'.

But Boyle is best remembered for his project to marry the philosophical endeavour of what he called 'the hermetick philosophers' to the practical and empirical pursuits of the 'vulgar spagyrists' (meaning chemists). In particular, he wanted to challenge unfounded assumptions about the nature and number of the elements. The philosophers espoused ancient Greek doctrines of the four elements, while the spagyrists believed in the *tria prima* of the Muslim alchemists and Paracelsus (see page 48).

Boyle wished to show that both were wrong, and in 1661 he published a dialogue, *The Sceptical Chymist, or Chymico-Physical Doubts & Paradoxes*, in which he dissected the flaws in their thinking and propounded his own view, which was that the true elements had yet to be discovered. He also gave the first clear definition of an element:

'certain primitive or simple, or perfectly unmingled bodies; which not being made of any other bodies, or of one another, are the ingredients of which all those called perfectly mixt bodies are immediately compounded, and into which they are ultimately resolved.'

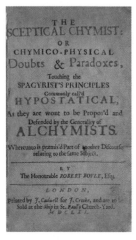

▲ Frontispiece of Boyle's *Sceptical Chymist*, one of the key texts of the Scientific Revolution.

1787

Year Guyton de Morveau et al. publish *Méthode de nomenclature chimique*

In 1787, French chemists Guyton de Morveau, Antoine Lavoisier, Antoine-François de Fourcroy and Claude-Louis Berthollet published *Méthode de nomenclature chimique* ('Method of Scientific Nomenclature'), which set out the first modern system of chemical nomenclature.

One legacy of chemistry's origins in alchemy and industry was a cacophony of names for substances, based on a mixture of principles: place of origin/extraction, colour, smell or taste, supposed magical or astrological qualities and associations, method of manufacture or local folklore and tradition. The same chemical might have different names in different languages and regions, and between different forms. Vague categories such as air, earth or oil had little or no scientific value.

In 1782, Guyton de Morveau suggested that the old names should be replaced with fixed, short names based on classical roots that would reflect the composition of a substance. The *Methode* of 1787 used Lavoisier's work as the basis for a new system. Elements would be given names, and these would be combined to produce the names of compounds. '*Une langue bien faite est une science bien faite*' (a well-made science depends on a well-made language), wrote Lavoisier, explaining in the preface to his later *Elements of Chemistry* (see page 136), 'we cannot ... improve a science without improving the language or nomenclature which belongs to it.'

1789

Year Lavoisier publishes the *Elements of Chemistry*

In 1789, in Paris, Antoine Lavoisier published his Traité Élémentaire de Chimie, known in English as the *Elements of Chemistry*. It was the most important textbook of scientific chemistry of its age – perhaps of any age.

Lavoisier was the son of a French lawyer who had trained in geology and mineralogy and fallen in love with science. He made a fortune by becoming a tax farmer and spent huge sums on the latest chemical equipment; he was assisted in his research and writing by his brilliant young wife, Marie-Anne (see page 71).

From the known to the unknown

An enthusiast for the English pioneers of the Scientific Revolution, he shared their philosophy of observation and experimentation before speculation. As he would later write, 'We ought to form no idea but what is a necessary consequence, and immediate effect, of an experiment or observation ... We should proceed from the known facts to the unknown.' Applying these principles to the gases discovered by Black and Priestley (see pages 36 and 46), Lavoisier dismantled the phlogiston theory and showed how combustion involved oxidation.

Off with his head

Lavoisier himself did not live to see the full fruit of his labours. Although he served the post-Revolutionary republic with diligence – for instance, helping to develop the metric system – his past as a tax farmer caught up with him. On 8 May 1794 he was executed by guillotine, his colleague Joseph-Louis Lagrange famously lamenting, 'It took only a moment to cut off that head, yet a hundred years may not give us another like it.'

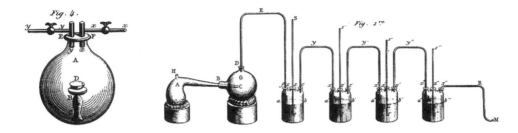

Developing Boyle's definition of an element (see page 134), he drew up a contingent list of elements and compounds, accepting that it was limited by the knowledge and technology available and predicting that his list would be modified and extended.

▲ Illustration from the *Elements*, engraved by Marie-Anne Lavoisier, who had trained herself in this art specifically to assist her husband (see page 71).

Transformed by degrees

In 1787 he collaborated with de Morveau and others in proposing a new system of scientific nomenclature (see page 135), which would both help to establish chemistry as a rigorous scientific discipline and enshrine his own theories about heat, combustion, compounds and acidity. Two years later, working on a follow-up to the earlier book, he found himself writing something longer, as he explained in the Preface to the *Elements*: 'Thus, while I thought myself employed only in forming a Nomenclature, and while I proposed to myself nothing more than to improve the chemical language, my work transformed itself by degrees, without my being able to prevent it, into a treatise upon the Elements of Chemistry.'

   The book that emerged proved to be the most influential chemistry text of its era, and perhaps in the whole history of chemistry. Not only did it lay the foundations of scientific chemistry, setting out with clarity principles and facts, it also inspired the subsequent history of chemistry. Men like Humphry Davy were profoundly influenced by the *Elements*, and the book set the agenda for the years that followed.

1808

Year Dalton's
New System published

John Dalton was a key figure in the history of chemistry and science in general, not just for his discoveries and theories but in his own person. Born in 1776 to a poor Quaker family in the provinces, he was very different to the gentleman amateurs who had mostly dominated British science until then, remaining outside the world of London's major scientific institutions. He is today remembered for his atomism or atomic theory and for his colour blindness.

Atomism

Atomism is the theory that matter is composed of indivisible particles, named 'atoms' by the ancient Greeks from their word for 'indivisible'. Revived by Boyle and Newton, atomism remained a theory of little practical value for chemists, who were unable to probe the properties of such infinitesimal particles. Dalton's breakthrough was to realise that atomism could help explain important findings in quantitative chemistry, settle one of its most contentious issues, and open the door to a profound new understanding of matter. At this time, there was a major debate over the make-up of compounds: were they made up of elements combined with each other in continuously variable proportions, or only in fixed ratios? Measurements of weights seemed to show that elements combined in simple integer ratios – e.g. 1 to 1, 2 to 1, etc. For Dalton, only atomism properly explained this, and in consequence elements must combine in fixed ratios.

Ultimate particles

Dalton set out his atomic theory in its fullest form in his 1808 *New System of Chemical Philosophy*, in which he defined the meaning of atomism for chemistry: 'The ultimate particles of all homogeneous bodies are perfectly alike in weight, figure, etc. In other words, every particle of water is like every other particle of water; every particle of hydrogen is like every other particle of hydrogen, etc.' These particles can neither be created nor destroyed, he insisted, adhering to the law of conservation of matter: 'No new creation or destruction of matter is within the reach of chemical agency. We might as well attempt to introduce a new planet into the solar system, or to annihilate one already in existence, as to create or destroy a particle of hydrogen.'

Atomic weights

Based on these atomic principles, Dalton was able to draw up a list of the known elements and assign to them relative atomic weights (now called relative atomic masses – see page 20). However, he assumed that nature is always parsimonious, so that the formula for water must be a single hydrogen atom bonded to a single oxygen atom. Taking the lightest element, hydrogen, as the reference point for 1, and knowing that in water the proportions by weight of hydrogen and oxygen are 1:8, he assigned an atomic weight of 8 to oxygen, using this in turn to define the weights of other elements but thus going off course by a factor of two. Equally problematic was his attempt to define a system of notation for the elements; Dalton used graphic symbols, arranged to show compounds in ways in close alignment with his theories. Perhaps more importantly, they were not easy for printers to use.

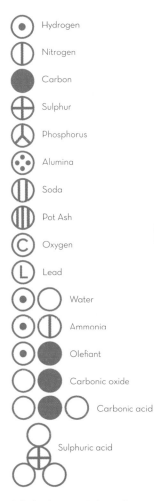

▲ Dalton's attempt at creating a system of universal scientific notation to overcome problems of language and custom. Difficult for typesetters, it also quickly became unwieldy for larger molecules, and never caught on (see page 148).

1828

Year Wöhler synthesises urea

In 1828, Friedrich Wöhler accidentally synthesised the organic chemical urea; this is often said to have marked the death knell for vitalism, the principle that organic chemistry depends on some special life force.

One of the great debates in the history of biochemistry was between vitalists and mechanists. Vitalists believed that living organisms had a special property or force that allowed them to make the complex molecules that characterised organic chemistry, so that such substances could not be synthesised in the lab. Mechanists believed that no processes are irreducibly complex, that life had no special properties or force and could be fully explained by chemical processes, and that all substances could theoretically be synthesised in the laboratory.

In 1828, not long after he had become the first to isolate aluminium from its compounds, German chemist Friedrich Wöhler was attempting to make ammonium cyanate. When he crystallised his preparation, he was astonished to find crystals identical to the organic urea crystals first isolated from urine in 1773. Writing to his mentor, Berzelius, he excitedly reported: 'I can no longer, so to speak, hold my chemical water and must tell you that I can make urea without needing a kidney, whether of man or dog; the ammonium salt of cyanic acid is urea.'

Wöhler's synthesis of urea is often said to have killed off vitalism at a stroke, but in fact the mechanists won the argument gradually over the next few decades.

1839

Year Charles Goodyear discovers vulcanisation of rubber

In the early 1830s, America was gripped by a rubber boom. But the natural product has serious drawbacks: being thermoplastic, it changes its properties with heat, becoming brittle in the cold and melting in the heat. In 1834, a hot summer finished off 'rubber fever', but a debt-ridden would-be inventor named Charles Goodyear determined to invent an improved rubber that would make his fortune.

After five years of failed experiments with a variety of additives, Goodyear was destitute and desperate. According to legend, his breakthrough came in 1839, when he was hawking his latest recipe, rubber gum with added sulphur. Met with derision at a general goods store, he angrily threw up his hands, accidentally hurled his sample onto the top of a sizzling hot stove and noticed that the heated mixture had a promising finish. Further experiments showed that rubber gum heated with sulphur has greater tensile strength, is harder to scratch and, crucially, remains elastic in heat or cold. In the process that came to be known as vulcanisation, after Vulcan, the Roman god of fire, long rubber polymers that normally lie parallel become cross-linked by bonding with sulphur, turning a mass of individual molecules into one giant, much more elastic molecule. Goodyear denied that his discovery was simply accidental, insisting that the incident with the hot stove only became significant to one 'whose mind was prepared to draw an inference', that is, the man who had 'applied himself most perseveringly to the subject'.

1856

Year William Henry Perkin synthesizes mauveine, the first synthetic dye

In 1856, British chemist William Henry Perkin accidentally synthesised the dye aniline purple, aka mauveine, kick-starting an industrial revolution in organic chemistry.

As the Industrial Revolution took off in the 19th century, industrial chemistry was in the vanguard, and the origins of what is now one of the biggest industries in the world lay in a garden hut in Stepney, east London. This was where an 18-year-old chemist, William Henry Perkin, pursued an accidental discovery he had made while trying to synthesise the anti-malarial drug quinine from aniline. Washing out a flask with alcohol after a failed attempt, Perkin found that the residue bled a startling mauve colour. Perkin and friends retreated to his garden shed at home, and here succeeded in making the first synthetic dye, aniline purple, marketed as mauveine.

Perkin's discovery was perfectly timed. The Industrial Revolution was in full swing, producing vast quantities of the coal tar he needed for raw material, developing garment factories for mass production, and creating a huge market for cheap and colourful clothes. Previous purple dyes had been prohibitively expensive (see page 153); mauveine caused a sensation, when Queen Victoria appeared at the Royal Exhibition of 1862 in a gown dyed with the substance. Perkin established a dye factory on the banks of the Grand Union Canal and developed many new colours, including Britannia Violet and Perkin's Green. It was said that the water in the canal turned a different colour every week.

1862

Year Kekule dreams up the structure of benzene

In 1862, according to his account of nearly 30 years later, as August Kekule dozed in an armchair before the fire, he dreamt of a snake, coiled in a ring, devouring its own tail; this dream revealed to him the ring structure of benzene.

Benzene is an organic molecule composed of six carbon atoms and six hydrogen atoms. It was discovered in 1825 by Michael Faraday and later isolated in great quantities from coal tar. As a building block for many important and useful organic chemicals, as well as a useful substance in its own right, benzene attracted much attention. Measurements of its molecular weight (78) led to the derivation of its formula (C_6H_6), but this left chemists with a great puzzle. The German chemist August Kekule had shown that carbon can form four bonds with other atoms, and in 1858 had achieved a great breakthrough, showing that carbohydrates are generally formed by chains of carbon atoms, with the ones not at the end of the chain using two of their four bonds to make the links in the chain. By this rule, a carbohydrate with 6 carbon atoms should have 14 hydrogen atoms projecting from them, and thus a molecular weight of 86, yet benzene does not.

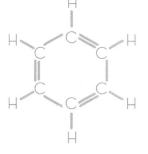

▲ The structure of benzene as worked out by organic chemists following Kekule, although because the actual arrangement of the bonds between the atoms in the ring is more fluid than suggested in this version, modern notation often simply uses a circle inside a hexagon to indicate a benzene ring.

The ourobouros ring

According to one of the most popular scientific legends since Newton's apple, the solution to this riddle came to Kekule in a dream. According to a story he told at a benzene symposium in

1890, he was gifted with two inspirational dreams. The first came to him around 1854, as he daydreamed aboard a horse-drawn London omnibus; he saw atoms dancing in chains, the smaller ones clustering at either end of a chain of larger ones. Kekule credited this dream for the inspiration for his 1858 insight into the structure of hydrocarbons. The second dream came as he slept in front of the fire in the winter of 1861–62. After seeing a dream vision of a snake devouring its own tail – linked by Carl Jung and others to an ancient alchemical symbol called the ourobouros, which symbolised the unity of nature – Kekule awoke with the realisation that in benzene the carbon atoms form a ring, like the ourobouros. This in turn meant that instead of forming single bonds with their neighbouring carbon atoms, three of the atoms form double bonds, thus leaving only six free bonds for hydrogen atoms. In his 1964 book on the creative process, *The Act of Creation*, Arthur Koestler called this 'probably the most important dream in history since Joseph's seven fat and seven lean cows'.

Dream cheat

In fact, Kekule's tale is highly suspect, and according to Dr John H. Wotiz, a professor of chemistry at Southern Illinois University, may have been a ploy to obscure his debt to chemists who had already worked out the ring structure of benzene: Archibald Scott Couper of Scotland, Joseph Loschmidt of Austria, and the French chemist Auguste Laurent, whose 1854 paper in the Paris journal *Méthode de Chimie* contained an illustration clearly showing the carbon atoms of benzene arranged in a hexagonal ring.

1869

Year Mendeleev
discovers the periodic law

In 1869, the Russian chemist Dmitri Mendeleev conceived the periodic law of chemistry, better known as the periodic table: the principle that governs how the elements assort into families that share physical and chemical properties.

The periodic law is arguably the greatest achievement and most important discovery in chemistry. It represented a grand synthesis of all the chemical knowledge accumulated up to that point, and helped explain many of the findings of the next century and a half. Mendeleev made this epochal breakthrough after sitting for three days at his table, arranging cards on which were written the names and atomic weights of every known element. On the third night he slept, and 'saw in a dream a table where all the elements fell into place as required. Awakening, I immediately wrote it down on a piece of paper.' The table he had sketched out showed that, arranged in rows and columns by atomic weight, the elements follow a periodic law, which is to say, their properties repeat at regular periods.

▲ Mendeelev, the grand old man of Russian chemistry, pictured in 1897 when he was 63 years old.

The telluric screw

Mendeleev was building on the work of others, most notably the French geologist Alexandre-Émile Béguyer de Chancourtois. After the Karlsruhe Conference of 1860 had accepted the theories of Avogadro (see page 77), it became possible to correct many errors in the atomic weights attributed to elements. Armed with this new information, de Chancourtois arranged the elements in order of their revised weights and spotted a pattern, which he called the *vis telluric*, or telluric screw. In his complex conception, the elements were ordered in helical form like the thread of a screw, with elemental properties repeating with a period of 16 atomic mass units. Unfortunately, de Chancourtois' background and obscure language, and the lack of an explanatory diagram in his paper, meant that he was destined for obscurity.

A suggested system

One man who had read the paper, however, was Mendeleev, a brilliant Russian chemist determined to discover 'the philosophical principles of our science which form its fundamental theme'. Working on a new textbook in 1869, Mendeleev began to muse on the family relations between the elements, noting at least three groups – the halogens and the oxygen and nitrogen families of elements – that could be arranged by atomic weight. With de Chancourtois in mind, he extended his purview to include all the elements, drawing up his cards and eventually dreaming the answer. In his 1869 paper 'A suggested system of the elements', Mendeleev presented a table at 90 degrees to the modern version, with elements arranged in columns of descending atomic weight, such that each row contained elements with similar properties. There were seven rows (the eighth, the noble gases, had yet to be discovered) and along each horizontal row the valencies of the elements went from 1 to 4 and back again to 1. Some of the elements in his table seemed out of place, whereas elsewhere there were gaps, which he predicted would be filled by new discoveries (see page 88), but the power of his vision commanded faith: 'Although I have had my doubts about some obscure points, yet I have never doubted the universality of this law, because it could not possibly be the result of chance.'

▼ Memorial to Mendeleev and his periodic table, in Bratislava, Slovakia.

1,913

Letters in the full chemical name for tryptophan synthetase

The full chemical name for the enzyme tryptophan synthetase runs to almost 2,000 letters, or at least it did until new naming conventions were introduced to cope with the length of complex organic chemicals.

Tryptophan synthetase is an enzyme: a biological catalyst made of protein that captures and guides reactants to a reactive site, facilitates steps in a reaction, and helps to distribute and mobilise the products. In the case of tryptophan synthetase, the enzyme catalyses the final step in the synthesis of tryptophan, an amino acid. Amino acids are the building blocks of proteins and in the 1960s there was a convention, when publishing research describing the structure of a protein, to give the protein a name that listed all its amino acids. Tryptophan synthetase is made up of 267 amino acids, so under this old convention its name included all 267 elements, starting off like this: methionyl**glutaminyl**arginyl**tyrosyl**glutamyl... (these are just the first 5 amino acids, alternately highlighted to show which is which). The end result is a name 1,913 letters long, for which space precludes the full inclusion.

Evidently this was unmanageable and not useful, so the convention was changed. It is likely that tryptophan synthetase was the largest protein thus named before this happened, and as a result it is often touted as the longest word in the English language. Today the systematic name of the enzyme is given as L-serine hydro-lyase.

2,000

Number of substances prepared, purified and analysed by Berzelius

Jöns Jakob Berzelius was an early 19th-century Swedish chemist who became the gatekeeper of his science for more than two decades, despite being on the fringes of Europe. He and his team discovered over 2,000 new elements and compounds.

In 1808, the same year that Dalton's *New System* came out, Berzelius published the first edition of his *Lärbok i Kemien*. Lately appointed a professor, Berzelius could find no suitable textbook in Swedish from which to teach, so he wrote his own. By force of character, hard work and experimental prowess, Berzelius would come to dominate European chemistry for the next two decades, his constant revisions of the *Lärbok* being widely translated and setting the agenda for the science across the continent. His most lasting contribution came after he had read Dalton's work and found the notation wanting (see page 139). During the preparation of his own book, Berzelius had also struggled with the messy nomenclature of chemistry, where the French system of Lavoisier et al was making slow headway amid a cacophony of linguistic and regional variation.

▲ Trained as a physician, Berzelius would become one of the founding fathers of scientific chemistry, after his genius for analysis was first spotted by a mine owner.

Latin notation

Seeking to solve both problems, Berzelius began to develop his own system of notation, taking his cue from his countryman Linnaeus, whose Latin-based system of nomenclature had set the standard for biology. Berzelius adopted or invented Latin names for every element he encountered, and instituted a system of notation using the initial letter or letters of the element. Thus sodium, isolated from soda, which was known to the ancients as natron, became natrium, with the notation Na. To iron he assigned the Latin name for the metal, ferrum, with the notation Fe. The notations for elements could then be combined to describe compounds, and Berzelius started the practice of using super- and subscripts for numbers, although he initially went for superscripts (so that sulphur dioxide became SO^2) and then used dots over the notation. The idea caught on, resulting in the modern use of subscripts for compounds. Ideas like this were circulated and propagated via his *Lärbok*.

Berzelius on bonds

As well as his role as a sort of clearing house and presiding authority for chemistry, Berzelius and his team made many important discoveries and first isolations, including many new elements such as cerium, selenium, thorium, lithium, silicon, vanadium and elements from the lanthanide series (see page 55).

His skill with fine measurements and quantitative analysis helped Berzelius to produce the most reliable table of atomic weights available at the time. Like Davy, he was quick to grasp the importance of the voltaic pile to chemistry, and on the back of his experiments in electrochemistry he formulated an influential theory making electrostatic attraction between negatively and positively charged atoms the basis for all chemical bonding. Made a baron in 1835, he died in 1848 after a long and fruitful career. Towards the end, however, as new discoveries and theories superseded his own contributions, Berzelius became increasingly sidelined and bitter.

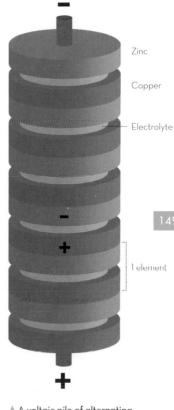

Zinc

Copper

Electrolyte

1 element

▲ A voltaic pile of alternating metal discs interleaved with pads soaked in electrolyte, such as Berzelius used in his researches.

3,550

Melting point of carbon, the highest of any element (°C)

The melting point of carbon in its graphite form, at relatively high pressure of 10 atmospheres, is 3,550°C (3,823 K). At normal atmospheric pressure, carbon won't melt at all but will change phase directly from solid to gas, known as sublimation.

Carbon has many unusual, even unique properties. It exists as a number of stable allotropes (different forms), including graphite, diamond, graphene, nanotubules and fullerenes or buckyballs. The most common by far is graphite. The physical properties of these allotropes span an incredible range. For instance, graphite is soft enough to use as a pencil (its name comes from the Latin for 'drawing'), while diamond is one of the hardest substances known in the universe (see page 64–65).

Carbon has the highest melting and boiling/sublimation points of any element. As graphene, carbon won't melt at normal atmospheric pressure, but will sublimate, which is where atoms migrate directly from solid to gas, at 3,825°C (4,098 K). Diamond has even higher melting/boiling points. Since the combustion temperature of diamond is around 800°C, a diamond would burn away long before you could evaporate it if heated in normal air. Diamond also has the highest thermal conductivity of any element, which means that it conducts heat better than any other element. Since a diamond at room temperature is below your body temperature, if you handle a diamond it will conduct heat away from your skin better than ice or cold metal, and will feel cool to touch. This is the origin of the slang term 'ice' for diamonds.

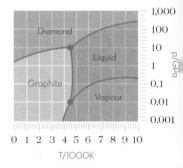

▲ Phase diagram of carbon, showing how the phases of different allotropes vary with temperature and pressure. At normal pressure (below the bottom of the y axis on this graph), the phase boundary for graphene is directly between solid and vapour; in other words graphene will sublimate, not melt, at high temperatures.

4,184

Joules in a kilocalorie

Lavoisier destroyed the phlogiston theory (see page 33) but in its place he promoted another delusory substance, caloric, to represent the principle of heat. Caloric, from the Latin word for 'heat', was supposed to be some sort of subtle fluid. Four years after Lavoisier's execution (see page 136), the metric system he had helped to develop was introduced in France, and soon after this, someone (it is not clear who) started to use the term 'calorie' for the amount of heat required to heat 1 kilogram of water by 1°C. Later, the meaning of calorie was changed to indicate the heat required to raise by 1°C a single gram of water, so the equivalent for a kilogram of water would be a kilocalorie (kcal). Even more confusingly, the kcal was rebranded in America as the Calorie.

Meanwhile, in Britain the chemist and brewer James Prescott Joule had shown through a series of elegant experiments that heat energy is interchangeable with mechanical and electrical energy, and that simple stirring could introduce heat to a closed system, helping to destroy caloric. He used units of pounds of water per degree Fahrenheit (still the definition of a British thermal unit or BTU), but later his name was given to the SI unit of energy, the joule (J), equal to the work done by a force of one newton acting through one metre, or 1 watt-second. One calorie equals 4.2 joules and a kilocalorie is equivalent to 4,184 joules.

Thermometer Stirrer

Solvent

Styrofoam coffee cups

▲ A simple way to recreate a calorimeter such as that used by Joule: Styrofoam cups insulate the system, while swizzling the stirrer adds energy, resulting in a temperature change that registers on the thermometer.

5,730

Half-life of the carbon
isotope ^{14}C (years)

The isotope of carbon with 14 protons in each atom is known as 14C or carbon-14 (C-14). It has a half-life of 5,730 years, and is widely used in the technique of carbon-dating.

There are three naturally occurring isotopes of carbon: ^{12}C, ^{13}C and ^{14}C. The latter makes up just 1 part per million of naturally occurring carbon. As a result, one CO_2 molecule in every million in the atmosphere will include an atom of carbon-14 rather than carbon-12, and because carbon gets into the food chain via plant photosynthesis, which involves fixing of atmospheric CO_2, living organisms exhibit in their tissues the same ratio of C-14 to C-12 as in the atmosphere. But C-14 is radioactive, with a half-life of 5,730 years, which means that of a given collection of C-14 molecules, about half will have decayed after this length of time. From the moment of death, the C-14 content of the organism starts to diminish as this radioactive isotope decays, and the previously constant ratio of C-14:C-12 in the organism's tissues will also start to diminish. Since the starting ratio is known to be roughly a million to one, measuring the current ratio in dead organic tissue provides a means to estimate how long ago it died.

This has proved to be a powerful tool for archaeologists, and carbon dating can be used to date organic remains up to 50,000 years old. Thus any wood, bone, antler, pollen, fabric, charcoal, pigment or other organic archaeological material can be directly dated, and associated finds, such as stone tools, can be dated by association.

12,000

Number of snails needed to produce 1.4g of Tyrian purple

Tyrian purple was a dark violet dyestuff produced since the Bronze Age and treasured by ancient peoples for its intense colour, resistance to fading and rarity. It is perhaps the most famous and fabled of pre-synthetic dyes, representing an era when the value of cloth could be determined by its colour.

The dye known as Tyrian purple was extracted from the hypobranchial gland of marine molluscs, most notably Murex species; 12,000 snails were needed to produce just 1.4 grams of the dye, making it extremely expensive.

Hercules' gift

According to one Greek legend, the dye got its name when Tyrus, a nymph beloved of Hercules, noticed that his dog had stained its mouth after chewing on a sea snail. She demanded a robe dyed in the same intense purple, leading Hercules to trawl the sea for thousands of molluscs and extract the first Tyrian purple. In fact, the dye got its name from the city of Tyre, one of the pre-eminent cities of the Phoenicians, a nation of seafaring traders. Tyre made its fortune off the back of what was, perhaps, history's first example of industrial chemistry: the extraction and production of Tyrian purple.

Blue gold

Other pre-synthetic dyestuffs were similarly difficult and expensive to make. Cochineal, made from the ground beetle, required 70,000 insects to produce a single pound (450 grams) of dye. Indigo, from the indigo plant, was known as 'blue gold' because it took 37 square metres worth of the crop to produce 100 grams of dye.

Dye hard

Although the precise secrets of the dye are lost to history, it probably involved fishing for two species of sea snail: the large *Murex trunculus* (or *brandaris*, sources differ) and the smaller *Buccinum lapillus*. The latter was simply crushed whole, but with the large *Murex* it was necessary to extract a mucus-filled tube, technically known as the hypobranchial gland. This was then subjected to lengthy processing, with Pliny describing how the extracted pulp was mixed with salt, soaked for three days, diluted with water and heated in lead cauldrons for ten more days while bits of mollusc flesh were skimmed off, until a clear liquid was left. According to the 1st century AD Greek geographer Strabo, the stench from the dye works made Tyre 'unpleasant for residence'. Some 3,600 kilograms of pulp produced just 225 kilograms of dye, according to one estimate. A secret process of blending and fixing led to a unique colour that darkened with age but would not fade.

Born in the purple

Tyrian purple was so expensive that the 4th-century BC Greek historian Theopompus reported that it 'fetched its weight in silver'. The city of Tyre grew rich on the dye industry, with the purple in great demand by aristocrats, kings and emperors. Eventually the Byzantine emperors forbade any but the imperial family to wear the dye, and the use of dyed cloths to wrap imperial newborns gave rise to the Greek term Porphyrogénnetos – 'born in the purple'.

23,000

Height to which Joseph Gay-Lussac ascended in a hydrogen balloon (feet)

On 16 September 1804, French chemist Joseph Gay-Lussac ascended alone in a hydrogen balloon to a height of 23,000 feet (7,016 metres), an altitude record that stood for nearly 50 years.

The discovery of hydrogen and the ability to liberate large amounts of this lightest and most buoyant of gases led to the balloon mania of the late 18th and early 19th centuries (see page 33). The inventor of the eponymous Charlier hydrogen balloon, Dr Charles, recklessly followed his accompanied initial flight by inviting his co-pilot to step out of the basket and ascending alone. Thus lightened, the balloon shot up rapidly to a height of 10,000 feet (3,050 metres) in just ten minutes, and Dr Charles was able to observe: 'I was the first man ever to see the sun set twice in the same day'.

▲ The product of two revolutions – the French and Chemical – Gay-Lussac was educated at the new Ecole Polytechnic, an institution set up to produce technocrats for the new order.

Up where the air is clear

The most scientifically productive of the early balloon flights was probably that of the 'wonder kid' of French chemistry, Joseph Gay-Lussac. After excelling at college, in 1801 Gay-Lussac had been tipped for greatness by Claude-Louis Berthollet, who is supposed to have said, 'Young man, it is your destiny to make discoveries.' In 1804, Berthollet sourced a hydrogen balloon from Egypt in order to help settle speculations about the relationship between altitude and the magnetic field of the Earth, and Gay-Lussac and a companion volunteered to ascend and take measurements. They took with them a menagerie of animals, releasing birds at various altitudes to observe their behaviour. Measurements of terrestrial magnetism proved inconclusive, so on 16 September Gay-Lussac went up again, on his own, reaching the unprecedented altitude of 20,000 feet (6,096 metres), where temperatures dropped to -9.5°C. Despite reaching the very limit of human survivability – 'my respiration was rendered sensibly difficult' – Gay-Lussac reported that 'I was far from experiencing any illness of a kind to make me descend.'

Combining volumes

Of greater significance than the measurements of magnetism he took, which appeared to show no appreciable diminution of the Earth's magnetic field with altitude, were Gay-Lussac's samples of air taken from different heights. Back in the lab, his analysis of these samples helped lead to some of his greatest discoveries, including what is now known as Gay-Lussac's Law, which states that at constant volume, pressure is directly proportional to temperature (see page 63). Perhaps more importantly, Gay-Lussac also helped elucidate the law of combining volumes, which says that in reactions between gases the ratios between the volumes of the reactants and products are simple integers. For instance, two volumes of hydrogen combine with one volume of oxygen to create one volume of water vapour. This would prove to be a vital piece in the puzzle of the nature of compounds and the atomic theory. It implied, for instance, that the correct formula for water is not HO – as Dalton had supposed (see page 76), with consequent errors in the relative atomic weights he assigned – but H_2O.

▼ Joseph Gay-Lussac and Jean-Baptiste Biot in their balloon, 24 August 1804, armed with instruments and animals.

400,000

Number of years ago that hominins achieved control of fire

The ancestors of modern humans probably made opportunistic use of fire from around 2 million years ago, but there is no clear evidence that early humans could maintain and control fire for domestic use until around 400,000 years ago.

There is some evidence that animals such as apes make use of the medicinal effects of certain plants, while elephants are said to seek out fermented fruits for their intoxicating effects. These might be said to represent examples of animal chemistry, and it is possible that our early hominid ancestors (members of the great ape group) also practised such rudimentary chemistry. However, the earliest hard evidence for human chemistry comes in the form of charred plants and bones associated with early hominins (members of the human family, including Australopithecines and *Homo* species), which appear to show use of fire. Found at the Wonderwerk Cave in South Africa, these charred remains occur alongside stone tools dated to 2 million years ago, which may have belonged to *Homo erectus*.

Better living through chemistry

This suggests that *Homo erectus* could take advantage of naturally occurring fires, and indeed humans evolved in a part of the world where natural sources of fire – such as lightning strikes and volcanic activity – are particularly plentiful. But true control of fire requires far greater cognitive and social skills than picking up a burning branch. 'Domestication' of fire came when humans could tend, maintain and, if necessary, transport fire, gathering and storing fuel and protecting the flame from the elements. When these skills were mastered, a better life through chemistry became available. Humans could cook their food, dramatically increasing the ratio of calories metabolised to calories expended in obtaining, preparing and digesting food. They could also extend their day, protect themselves from predators, survive cooler temperatures, harden spear points and start to experiment with materials such as heat-treated clay, glue and pigment.

Strike a light

The ability to start fires from scratch, known as ignition, seems not to have developed until the Upper Palaeolithic era, with the earliest potential 'strike-a-light' tool made from pyrites dating to 80–40,000 years ago. Fire-lighting may have been an achievement reserved exclusively for *Homo sapiens*.

Neanderthal fire

When did humans discover how to control fire? Some palaeoanthropologists interpret evidence from the Israeli site Gesher Benot Ya'aqov as demonstrating controlled use of fire, probably by *Homo erectus*, but this is not widely agreed. According to palaeoanthropologists Wil Roebroeks and Paola Villa, it is not until the Middle Pleistocene, around 400–200,000 years ago, that there was 'habitual use of fire'. This is surprising, because hominins had already been occupying Ice Age Eurasia for around 600,000 years by this time. Roebroeks and Villa conclude that it is only 'with the Neanderthals and their contemporaries elsewhere in the Old World, that fire became an integral part of the technological repertoire of the human lineage.' In other words, the Neanderthals may have been the first hominins to control fire.

570,000

Minimum molecular
weight of cellulose (g/mol)

The most abundant naturally occurring organic compound is cellulose, a polymer composed of glucose units, combined in very long chains with massive molecular weights of at least 570,000 grams per mole.

Cellulose comprises up to 33 per cent of plant matter (up to 90 per cent in cotton). It is very strong and not soluble in water; plants use it as their main structural component, to provide strength and stability. Yet it is made from the exact same ingredient as starch, which is water soluble and weak.

Both starch and cellulose are composed of linked glucose units. Glucose is a saccharide, hence cellulose is a polysaccharide. In starch, the glucose units are all oriented in the same direction, but in cellulose every other glucose is reversed 180 degrees. In cellulose, the chain of glucose can be anywhere from a few hundred units long in wood pulp to well over 6,000 in cotton fibres, reaching up to 0.04 millimetres long. The hydroxyl groups that project from the glucose chain can form hydrogen bonds with groups on other chains, and the way that the units are oriented in cellulose means that chains can stack together in places to form hard, stable, crystalline regions; these two features combine to give cellulose its strength and insolubility.

Although humans lack the digestive enzymes to break down cellulose, it is useful both as wood and cotton products (including paper), and in processed form as the basis for polymers such as cellulose acetate (aka celluloid for films) and rayon (cellulose xanthate).

40 million

Energy density of a ton of uranium (kWh)

A ton of natural (non-enriched) uranium can theoretically produce 40 million kilowatt-hours of energy. This is equivalent to burning 16,000 metric tons of coal or 80,000 barrels of oil.

The energy density of a material is a measure of how much energy it produces per unit of mass. Of all the raw materials used in power generation, uranium has by far the greatest energy density. To understand the potential value of the high energy density of uranium, it is necessary to compare it to other sources of energy. A kilogram of wood provides roughly 1 kWh or 10 megajoules (MJ) of energy, enough to power a 100W light bulb for 1.2 days. A kilogram of coal could power the light bulb for 3.8 days, diesel for 5.3 days. Natural uranium has an energy density of 57,000 MJ/kg, and 1 kilogram could power the bulb for 182 years in a traditional nuclear reactor, or 25,700 years in a 'breeder' reactor.

Amount of fuel needed to run a 1-gigawatt power station for a year for different materials, according to the International Atomic Energy Agency (IAEA):

Fuel	Mass (metric tons)	Equivalent
Coal	2.6 million	2,000 train carriage loads
Oil	2 million	10 supertankers
Uranium	30	Reactor core measuring 10 m on each side

90 million

90 million and counting: number of organic and inorganic substances registered with CAS

The Chemical Abstracts Service (CAS), a division of the American Chemical Society, maintains the CAS Registry, a list of every publicly disclosed substance. To date it has registered over 90 million unique substances.

The CAS Registry compiles data from patents, journals, chemical catalogues and some online sources to record molecular structure, chemical name, molecular formula and other identifying information, every time a new substance is encountered. The substance is assigned a CAS Registry Number, which acts as its globally recognised identifier. The registry comprises organic and inorganic substances, alloys, coordination compounds, minerals, mixtures, polymers and salts.

A phenomenal 15,000 new substances are added to the Registry every day: this is one every 6 seconds. So by the time you finish reading this entry, there will be 10–15 new substances that have come into 'official' existence. The number of possible chemicals is infinite. For example, in 2013 researchers at Duke University produced a 'map' of the small-molecule universe – the 'chemical space covering all possible small organic molecules' – that included more than 10^{60} molecules; this is just for organic molecules with a molecular mass of 500 Da or less. In other words, the CAS Registry will not be running out of candidates any time soon.

200 million

Molecular mass of PG5, largest synthetic molecule ever created (amu)

The dalton is a unit of atomic mass named after the father of modern atomic theory, British scientist John Dalton. One dalton is the same as 1 amu (see page 20), and the latter unit is now preferred. However, daltons are still used in many fields, particularly megamolecule chemistry, in which the molecular masses of giant molecules are given in megadaltons or MDa – millions of daltons.

   In nature, very large or 'macromolecules' are common, such as plant polymers (e.g. cellulose – see page 159). Humans have learned to synthesise their own massive polymers, such as nylon (see page 56). Until 2011, the largest stable synthetic molecule was polystyrene, a polymer of which could be equivalent in mass to 40 million hydrogen atoms (40 MDa). But in that year, a team from the Swiss Federal Institute of Technology in Zurich published research on PG5, the largest stable synthetic molecule ever created, with a molecular mass of 200 million daltons or amu, equivalent to the mass of 200 million hydrogen atoms. The team used polymerisation methods to create a backbone, to which they added radiating groups of atoms. There is a school of thought that any cross-linked polymers such as rubber or carbon fibre constitute single molecules, however, in which case the largest molecule might be a giant tyre or the hull of a ship. Diamonds can also be considered single molecules, in which case the crystallised white dwarf star BPM 37093, effectively a single diamond, may be the largest single molecule known.

3.2 billion

Number of bases
in human DNA

DNA is made up of a string of subunits called nucleotides, which contain nitrogenous compounds called bases. Each base makes up a letter in the DNA code, and the human genome is 3.2 billion bases long.

In 1869, Swiss biochemist Friedrich Miescher first isolated DNA, which he called nuclein, from the large nuclei of white blood cells, obtaining large quantities of these by collecting pus-soaked bandages from a clinic.

In 1879, German biologist Walther Flemming observed the presence and duplication during cell division of chromosomes, filaments in the nucleus of a cell, and analysis showed that chromosomes contain both DNA and protein. Attention focused on protein as the likely agent of heredity, as proteins were known to be complex, while chemical analysis of DNA seemed to suggest a relatively simple organisation, composed of units called nucleotides, each comprising a phosphate, a sugar called deoxyribose, and four bases called adenine, cytosine, guanine and thymine.

In 1944, however, American microbiologist Oswald Avery and his team proved that DNA is the agent of genetic transmission, firing the starting gun on a race to work out the structure of DNA and unravel the genetic code. The race was famously won by English biophysicist Francis Crick and young American microbiologist James Watson, at Cambridge University's Cavendish Laboratory, in 1953. They showed how DNA forms a double helix with two strands connected by hydrogen bonding between base pairs, with the sequence of the pairs constituting the genetic code.

9,192,631,770

Microwave frequency of
a caesium atom (Hz)

The SI definition of a second is the time taken for the microwave radiation emitted by the outer electron of a caesium atom as it transitions between states to complete 9,192,631,770 full waves. The microwave radiation emitted by this transition thus has a frequency of 9,192,631,770 Hz.

Prior to 1967, the definition of a second was based on the orbital period of the Earth, but this is too inaccurate for measurements of general relativity and the running of the GPS system. A natural phenomenon that never changes its properties and is always the same is the amount of energy emitted when an electron transitions from a higher energy state to a lower one. This energy takes the form of electromagnetic waves, with frequencies that can be determined and used to provide the ultimate reference for a system of timekeeping.

The isotope most widely used for these 'atomic clocks' is caesium-133, element number 55. Caesium is preferred because 54 of its electrons occupy completely filled orbitals and thus form a very stable symmetric core, with almost no interaction with the 55th, outermost electron. This makes it easier to observe precisely the transitions of the outermost electron, as its emissions give a very narrow spectral line the wavelength or frequency of which can be accurately determined. Since 2013, the caesium clock used by the US National Institute of Standards and Technology has an uncertainty of just 3×10^{-16}, equating to an accuracy of 1 second in over 100 million years.

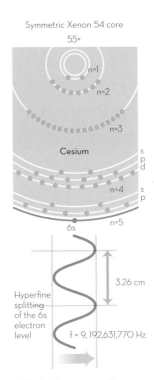

▲ Illustrative cross-section through a caesium atom, showing the concentric electron shells and the 'orphan' outermost electron generating microwaves of very precise frequency.

3.7×10^{10}

Number of atoms in 1g of radium-226 that will decay in 1 second, aka 1 curie

The definition of a curie, a unit of radioactivity, is the amount of material that will produce 3.7×10^{10} nuclear decays per second, which in turn is derived from the number of atoms in a gram of radium-226 that decay in a second.

The curie (Ci) is just one of a host of competing and complementary units used to measure radiation, the dose absorbed by a human and the biological effect of the dose absorbed. Each of these quantities has a traditional, popular unit and a newer SI unit. The curie was based on the radioactivity of radium, the element discovered by Marie Curie (see page 40). Its SI replacement, the becquerel (Bq), named for Henri Becquerel, the discoverer of radioactivity, is one decay per second. Hence 1 Ci = 37 billion Bq.

The curie and becquerel measure the radioactivity of a material, but don't say anything about the type of radiation or, more importantly, its effect on humans, which depends on the amount absorbed, and on the type and energy of that radiation. The former was traditionally measured with the rad, defined as an absorbed dose of 0.01 joules of energy per kilogram of body tissue, but the SI unit is now the gray (Gy), defined as 1 joule of deposited energy per kilogram of tissue, and hence equivalent to 100 rads. The biologically effective dose was measured in rems (which stands for Roentgen Equivalent Man), still commonly used in the US, but the SI unit is now the sievert (Sv), which equals 100 rems.

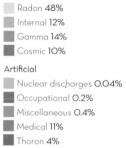

Natural
- Radon 48%
- Internal 12%
- Gamma 14%
- Cosmic 10%

Artificial
- Nuclear discharges 0.04%
- Occupational 0.2%
- Miscellaneous 0.4%
- Medical 11%
- Thoron 4%

▲ Pie chart showing the sources of radiation to which an average person is exposed in their lifetime. By far the greatest proportion comes from natural sources, mostly from naturally occurring radon gas.

1.7588196×10^{11}

Charge/mass ratio of the electron (coulomb/kg)

The ratio of the charge on the electron (measured in coulombs, C) to the mass of the electron (measured in the SI unit the kilogram, kg) is 1.7588196×10^{11} C/kg. The determination of this ratio was one of the great discoveries in the history of science, proving the existence of subatomic particles and finally disproving the millennia-old atomic creed of the ancient Greeks.

So promising an opportunity

The electron is said to have been discovered in 1897 by English physicist J.J. Thomson. What he actually did was to determine accurately the charge/mass ratio of the unknown particles that made up cathode rays. Cathode rays were discovered thanks to the Crookes tube and similar devices: mostly evacuated sealed tubes, from which almost all of the air had been evacuated to give a partial vacuum, contained the negative terminal – cathode – of an electric circuit. When the current is turned on, a beam or ray is projected from this terminal; accordingly it was named the cathode ray. Experiments showed that the cathode ray could be deflected by magnetic fields and could produce phenomena associated with electromagnetism, such as producing light and X-rays; they became a subject of great interest for scientists. 'There is no other branch of physics,' wrote Thomson in 1893, 'which affords us so promising an opportunity of penetrating the secret of electricity.'

▲ J.J. Thomson won the 1906 Nobel Prize in Physics. His 'corpuscle' quickly became known as the electron, and he himself presciently suggested that it might explain the nature of the periodic table.

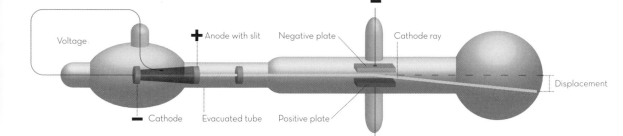

Voltage

+ Anode with slit   Negative plate   Cathode ray

Displacement

Cathode   Evacuated tube   Positive plate

A startling idea

Although German scientists led by Heinrich Hertz believed the cathode ray to be a wave, like light, Thomson was sure it was a beam of particles, which he referred to as corpuscles. The only question was: what sort of particles? Were they atoms, or something smaller than the supposedly indivisible atom? Hertz's work provided a strong clue as to the latter, for the German had found that cathode rays could penetrate through gold film. 'The idea of particles as large as the molecules of a gas passing through a solid plate was a somewhat startling one,' Thomson later observed; they must be smaller. His own experiment, in which he gave a metal bucket a negative charge by firing into it a cathode ray, showed that the particles were charged.

    Measuring the mass of such a tiny particle directly would be impossible, but if he could measure the ratio between its mass and its charge he could infer the former from the latter. Thomson knew this ratio could be calculated by finding the velocity of the particles, and that this in turn could be determined by subjecting the particles to balanced electrical and magnetic fields and noting the ratio of field strengths. He devised a modified Crookes tube to achieve this, finding that 'for the corpuscle in the cathode rays the value of charge/mass is 1,700 times the value for the corresponding quantity for the hydrogen atom'. 'We are driven to the conclusion,' Thomson said in his lecture that accompanied his receipt of the 1906 Nobel Prize in Physics, 'that the mass of the corpuscle is only about $1/1,700$ of that of the hydrogen atom. Thus the atom is not the ultimate limit to the subdivision of matter.' Thomson's 'corpuscle' quickly became known as the electron, and he himself suggested that it might explain the nature of the periodic table.

▲ The basic set-up of Thomson's landmark experiment, with a modified Crookes tube generating a cathode ray, a narrow beam of which is admitted through a slit in the anode. As it passes between charged plates and electromagnetic coils (not shown) the effects of the fields can be calculated from the degree of displacement measured on the fluorescent screen coating the bulb at right.

6.0221367×10^{23}

Avogadro's number

Avogadro's number or constant, N_A, is defined as the number of atoms in exactly 12 grams of carbon-12, and has been experimentally determined as 6.0221367×10^{23}, or roughly 600 billion trillion.

Named for the Italian lawyer and scientist Amedeo Avogadro, the Avogadro number is a vital constant for chemistry as it allows chemists to relate atomic mass to measurable mass in the real world, via the concept of the mole. A mole is the amount of a substance that contains Avogadro's number of particles. The particles can be anything from photons or electrons to ions, atoms or molecules.

The power of the mole

The concept of the mole originates with Avogadro himself, who devised the ideal gas law known as Avogadro's Law, which states that equal volumes of different gases, at the same temperature and pressure, contain an equal number of molecules. Although not appreciated at the time, this proved to be an immensely powerful idea because it provides a simple way to measure – in relative terms – how many atoms or molecules of a substance are present. Counting atoms or molecules is impossible, but the mole provides a means to count by weighing, since the definition of the mole means that an Avogadro's number of particles weighs the same in grams as the atomic weight of the particle.

Thus a mole of carbon-12 weighs 12 grams. If you have 48 grams of carbon-12, you know that you have 4 moles of it, and if you find that this amount of carbon reacts with 64 grams of O_2 gas (aka 2 moles), you can deduce that the resulting gas is carbon monoxide, not carbon dioxide. Thus the mole concept makes it possible for analytical chemists to work out the empirical formula of the product of a reaction by weighing the reactants.

Counting the uncountable

Avogadro himself had no way of working out the number that now bears his name, and it was first given the name by French physicist Jean Perrin in 1908. By this time there had already been various attempts to work out the magnitude of N_A, starting with Josef Loschmidt, an Austrian chemistry teacher, in 1865; accordingly, it was initially known as Loschmidt's number. James Clerk Maxwell estimated that in a cubic centimetre of gas at STP, Loschmidt's number was 1.9×10^{19}, which equates to an N_A of 4.3×10^{23}. Lord Kelvin used gaseous light scattering to arrive at an estimate of 5×10^{23}, while Perrin used studies of Brownian motion (where microscopic particles are buffeted by air molecules) to estimate the number to be 6.5–6.9×10^{23}. Today, X-ray crystallography allows very precise determination of the number, which is now known with an error of less than 0.00000001.

How big is that?

Avogadro's number is the biggest one in this book. It is so astronomical that it is hard to grasp. For instance, it is about 100 trillion times bigger than the population of the Earth. The vast size of this number underlines the infinitesimal size of atoms. A mole of water (18 grams) can be contained in a large tablespoon. A mole of fine sand would cover the whole of Texas in a layer 15 metres deep.

Further reading

Useful websites

American Chemical Society acs.org

American Institute of Physics aip.org

American Physical Society aps.org

Bureau International des Poids et Mesures bipm.org

Chemguide chemguide.co.uk

Chemical and Engineering News cen.acs.org

Chemical Heritage Foundation chemheritage.org

Chemicool Periodic Table chemicool.com

ChemWiki: The Dynamic Chemistry E-textbook chemwiki.ucdavis.edu

Classic Papers from the History of Chemistry web.lemoyne.edu/giunta

CO_2now.org Earth's CO_2 Home Page: co2now.org

Graphene Manchester's Revolutionary 2D Material: graphene.manchester.ac.uk

International Atomic Energy Agency iaea.org

International Commission on the History of Meteorology meteohistory.org

International Energy Agency iea.org

Jefferson Lab, It's Elemental Periodic Table: education.jlab.org/itselemental/

National Centre for Biotechnology Information ncbi.nlm.nih.gov

National Institute of Standards and Technology nist.gov

National Oceanic and Atmospheric Administration noaa.gov

Oak Ridge National Laboratory ornl.gov

Office of Science and Technical Information osti.gov

Perspectives on Plasmas plasmas.org

Phys.org phys.org

Royal Society of Chemistry, Periodic Table www.rsc.org/periodic-table

Royal Society of Chemistry rsc.org

Science is Fun in the Lab of Shakhashiri scifun.chem.wisc.edu

Scripps Institution of Oceanography scripps.ucsd.edu

Steve Lower's General Chemistry virtual textbook chem1.com/acad/webtext/virtualtextbook.html

The Chem Team chemteam.info

The Discovery of the Electron, American Institute of Physics, 1997 A Look Inside the Atom aip.org/history/electron/jjhome.htm

The International Association for the Properties of Water and Steam iapws.org

U.S. Energy Information Administration eia.gov

U.S. Geological Survey usgs.gov

WebElements periodic table webelements.com

Woods Hole Oceanographic Institute whoi.edu

Useful periodicals

American Scientist

APS News

Chemical Heritage Magazine

Chemistry World

Education in Chemistry

Nature

New Scientist

Physics Today

Popular Mechanics

Popular Science

Selected books

Ball, Philip, *The Elements: A Very Short Introduction*; Oxford: Oxford Universtiy Press, 2004

Brock, William H., *The Norton History of Chemistry*; W. W. Norton & Co. Inc., London: 1993

Brock, William Hodson, *William Crookes (1832-1919) and the Commercialisation of Science*, Aldershot. Ashgate, 2008

Brown, Theodore L., Lemay Jr, H. Eugene, Bursten, Bruce E., Murphy, Catherine J., Woodward, Patrick M., Stoltzfus, Matthew W., *Chemistry: The Central Science*; Prentice Hall, 13th edition, 2014

Crone, Hugh D., *Paracelsus, the Man who Defied Medicine: His Real Contribution to Medicine and Science*; Albarello Press, 2004

Ebbing, Darrell, Gammon, Steven D., *General Chemistry, Enhanced Edition*; Independence, Kentucky: Cengage Learning, 2010

Emsley, John, *The Elements of Murder*; Oxford: Oxford University Press, 2006

Emsley, John, *The Elements*; 3rd Edition, Oxford: Clarendon Press, 1998

Fara, Patricia, *An Entertainment for Angels: Electricity in the Enlightenment*; Cambridge: Icon, 2003

Gratzer, Walter, *Eurekas and Euphorias: The Oxford Book of Scientific Anecdotes*; Oxford: Oxford University Press, 2002

Hartston, William, *The Book of Numbers*; London: Metro, 2000

Heilbron, J. L., ed., *The Oxford Companion to the History of Modern Science*; Oxford: Oxford University Press 2003

Hermes, Matthew; *Enough for One Lifetime: Wallace Carothers, Inventor of Nylon*; Philadelphia: Chemical Heritage Foundation, 1996

Holmes, Richard, *The Age of Wonder*; London: HarperCollins, 2008

Hunter, Graeme K., *Vital Forces: The Discovery of the Molecular Basis of Life*; Academic Press, 2000

Jay, Mike, *The Atmosphere of Heaven: The Unnatural Experiments of Dr Beddoes and His Sons of Genius*; New Haven, Conn.; London: Yale University Press, 2009

Johnson, Stephen, *The Invention of Air: An experiment, a journey, a new country and the amazing force of scientific discovery*; London: Penguin, 2009

Judson, Horace Freeland, *The Eighth Day of Creation: Makers of the Revolution in Biology*; New York: Cold Spring Harbor Laboratory Press, 1996

Kelly, Jack (2004), *Gunpowder: Alchemy, Bombards & Pyrotechnics: The History of the Explosive that Changed the World*; New York: Basic Books

Kinger, Thomas B., ed.; *Nuclear Energy Encyclopedia: Science, Technology, and Applications*; Chichester, Sussex: John Wiley & Sons, 2011

Klaasen, Curtis D., *Casarett & Doull's Toxicology: The basic science of poisons*, 7th Edition; McGraw Hill Medical, 2008

Knight, David, *Ideas in Chemistry: A History of the Science*; New Brunswick: Rutgers University Press, 1992.

Levy, Joel, *A Bee in a Cathedral*; Richmond Hill, Ontario: Firefly Books, 2011

Levy, Joel, *Poison: A Social History;* Stroud, Gloucestershire: The History Press, 2011

Levy, Joel, *The Bedside Book of Chemistry*; Sydney: Murdoch Books Pty Ltd, 2011

Malley, Marjorie Caroline, *Radioactivity: A History of a Mysterious Science;* Oxford: Oxford University Press, 2011

Meli, Domenico Bertoloni, *Thinking with Objects: The Transformation of Mechanics in the Seventeenth Century;* Baltimore: The Johns Hopkins University Press, 2006

Pais, Abraham, *Inward Bound: Of Matter and Forces in the Physical World*; Oxford: Clarendon Press, 1986

Principe, Lawrence M., *The Aspiring Adept: Robert Boyle and his Alchemical Quest*; Princeton: Princeton University Press, 1998

Rhodes, Richard, *The Making of the Atomic Bomb*; London: Penguin, 1988

Timbrell, John, *The Poison Paradox*; Oxford: Oxford University Press, 2008

Week, Andrew, *Paracelsus: Speculative Theory and the Crisis of the Early Reformation*; Albany: State University of New York Press, 1997

Wrangham, Richard, *Catching Fire: How Cooking Made Us Human*; New York: Basic Books, 2009

Index

172

175

Picture credits